Fault Diagnosis for Electric Power Systems and Electric Vehicles

The present monograph offers a detailed and in-depth analysis of the topic of fault diagnosis for electric power systems and electric vehicles. First, the monograph treats the problem of *Fault diagnosis with model-based and model-free techniques (Model-based fault diagnosis techniques and Model-free fault diagnosis techniques)*. Next, the monograph provides a solution for the problem of *Control and fault diagnosis for Synchronous Generator-based renewable energy systems (Control of the marine-turbine and synchronous-generator unit and Fault diagnosis of the marine turbine and synchronous-generator unit.* Additionally, the monograph introduces novel solutions for the problem of *Fault diagnosis for electricity microgrids and gas processing units (Fault diagnosis for electric power DC microgrids and Fault diagnosis for electrically actuated gas compressors)*. Furthermore, the monograph analyzes and solves the problem of *Fault diagnosis for gas and steam-turbine power generation units (Fault diagnosis for the gas-turbine and Synchronous Generator electric power unit and for the steam-turbine and synchronous generator power unit)*. Finally, the monograph provides a solution for the problem of *Fault diagnosis for wind power units and for the distribution grid (Fault diagnosis for wind power generators and Fault diagnosis for the electric power distribution grid)*.

- The new fault detection and isolation methods that the monograph develops are of generic use and are addressed to a wide class of nonlinear dynamical systems, with emphasis on electric power systems and electric vehicles.
- On the one side, model-based fault detection and isolation methods are analyzed. In this case, known models about the dynamics of the monitored system are used by nonlinear state observers and Kalman Filters, which emulate the system's fault-free condition.
- On the other side, model-free fault detection and isolation methods are analyzed. In this case, raw data are processed by neural networks and nonlinear regressors to generate models that emulate the fault-free condition of the monitored system.
- Statistical tests based on the processing of the residuals, which are formed between the outputs of the monitored system and the outputs of the fault-free model provide objective and almost infallible criteria about the occurrence of failures.
- The new fault detection and isolation methods with statistical procedures for defining fault thresholds enable early fault diagnosis and reveal incipient changes in the parameters of the monitored systems.

Fault Diagnosis for Electric Power Systems and Electric Vehicles

Gerasimos Rigatos
Masoud Abbaszadeh
Mohamed-Assaad Hamida
Pierluigi Siano

CRC Press

Taylor & Francis Group

Boca Raton London New York

CRC Press is an imprint of the
Taylor & Francis Group, an **informa** business

First edition published 2025
by CRC Press
2385 NW Executive Center Drive, Suite 320, Boca Raton FL 33431

and by CRC Press
4 Park Square, Milton Park, Abingdon, Oxon, OX14 4RN

CRC Press is an imprint of Taylor & Francis Group, LLC

ISBN: 978-1-032-86451-8 (hbk)
ISBN: 978-1-032-86465-5 (pbk)
ISBN: 978-1-003-52765-7 (ebk)

DOI: 10.1201/9781003527657

Typeset in Nimbus Roman
by KnowledgeWorks Global Ltd

Publisher's note: This book has been prepared from camera-ready copy provided by the authors.

Contents

Preface ix

Author xi

Chapter 1   Fault diagnosis with model-based and model-free techniques 1

    1.1   Model-based fault diagnosis 1
        1.1.1   Outline 1
        1.1.2   Dynamic model of the three-phase inverter 2
        1.1.3   Approximate linearization of the three-phase
                 inverter 6
        1.1.4   Statistical fault diagnosis with the H-infinity
                 Kalman Filter 9
        1.1.5   Simulation tests 12
    1.2   Model-free fault diagnosis 16
        1.2.1   Outline 16
        1.2.2   Dynamic model of the PMLSM 19
        1.2.3   Neural-network model of the PMLSM 21
        1.2.4   Statistical fault diagnosis using the neural network. 26
        1.2.5   Simulation tests 30

Chapter 2   Fault diagnosis for SG-based renewable energy systems 37

    2.1   Control of marine-turbine and SG 37
        2.1.1   Outline 37
        2.1.2   Dynamic model of the marine-turbine power unit ... 40
        2.1.3   Nonlinear optimal control for marine turbine
                 with SG 42
        2.1.4   Nonlinear optimal control of marine turbine
                 and SG 47
        2.1.5   Flatness-based control for marine-turbine and SG .. 50
        2.1.6   Multi-loop flatness-based control for
                 marine-turbine and SG 53
        2.1.7   Sliding-mode control for marine-turbine and SG 56
        2.1.8   Multi-model fuzzy control for marine-turbine
                 and SG 59
        2.1.9   Simulation tests 61
    2.2   Fault diagnosis of marine-turbine and SG 82
        2.2.1   Outline 82

v

| | 2.2.2 | Differential flatness for marine-turbine and SG power units | 84 |

2.2.3   Design of a Kalman Filter-based disturbance observer ... 85

2.2.4   Statistical tests for fault detection and isolation ... 86

2.2.5   Simulation tests ... 88

Chapter 3   Fault diagnosis for electricity microgrids and gas processing units ... 98

3.1   Fault diagnosis of DC microgrids ... 98

     3.1.1   Outline ... 98

     3.1.2   Dynamic model of the DC microgrids ... 100

     3.1.3   Linearization of the state-space model of DC microgrids ... 102

     3.1.4   Statistical fault diagnosis with the H-infinity Kalman Filter ... 104

     3.1.5   Simulation tests ... 108

3.2   Fault diagnosis for 5-phase IM-driven gas compressors ... 113

     3.2.1   Outline ... 113

     3.2.2   Dynamic model of the 5-phase IM-driven gas compressor ... 114

     3.2.3   Linearization of the 5-phase IM-driven gas-compressor ... 118

     3.2.4   State estimation and fault diagnosis ... 122

     3.2.5   Simulation tests ... 126

Chapter 4   Fault diagnosis for gas and steam-turbine power generation units ... 139

4.1   Fault diagnosis of gas-turbine and SG ... 139

     4.1.1   Outline ... 139

     4.1.2   Dynamic model of the gas-turbine power unit ... 140

     4.1.3   Differential flatness of the gas-turbine power unit. 146

     4.1.4   Input-output linearization of the power unit ... 148

     4.1.5   State estimation with nonlinear Kalman Filtering. 153

     4.1.6   Statistical fault diagnosis using the Kalman Filter 155

     4.1.7   Simulation tests ... 157

4.2   Fault diagnosis of steam-turbine and SG ... 167

     4.2.1   Outline ... 167

     4.2.2   Dynamic model of the steam-turbine and SG ... 169

     4.2.3   Differential flatness of the steam-turbine and SG.. 172

     4.2.4   Differential flatness theory-based Kalman Filtering ... 177

     4.2.5   Statistical fault diagnosis with the Kalman Filter .. 180

     4.2.6   Simulation tests ... 182

Chapter 5 Fault diagnosis for wind power units and the distribution grid .. 191

5.1 Fault diagnosis for wind-power generators 191
  5.1.1 Outline.. 191
  5.1.2 Dynamic model of the wind power unit 192
  5.1.3 Differential flatness of the wind-turbine and SG ... 196
  5.1.4 Input-output linearization of the wind power unit . 197
  5.1.5 State estimation with Kalman Filtering 198
  5.1.6 Simulation tests.. 201
5.2 Fault diagnosis for the electric grid 201
  5.2.1 Introduction.. 201
  5.2.2 Model of the power grid 204
  5.2.3 Sensors' condition monitoring with Kalman
         Filtering.. 205
  5.2.4 Fault detection with the use of statistical criteria... 206
  5.2.5 Use of the Kalman Filter as a disturbance
         observer.. 208
  5.2.6 Simulation tests.. 210

References **215**

Index 233

Preface

The aim of this monograph on "Fault Diagnosis for Electric Power Systems and Electric Vehicles" is to analyze methods for fault detection and isolation in electric power systems and in electric traction and propulsion systems. To this end, the following topics are treated:

(a) model-based fault diagnosis with the use of nonlinear estimation methods and statistical fault diagnosis criteria. Modeling of the fault-free condition of nonlinear systems with the use nonlinear Kalman filters and nonlinear observers are extensively presented. At a next stage statistical decision making for fault detection and isolation is analyzed. Statistical approaches to fault diagnosis are introduced as methods that allow for precise and optimized definition of fault thresholds and which permit for distinguishing for incipient parametric changes in the monitored systems and for minimizing the launching of false alarms. In this context, an approach for fault thresholds definition based on the confidence intervals of the χ^2 (chi-square) distribution is presented. Several application examples are given about fault detection and isolation in the electric power transmission and distribution system, interconnected electric power units, conventional AC power units, renewable energy sources, electric power microgrids, and electric machines.

(b) model-free fault diagnosis with the use of nonlinear estimation methods and statistical fault diagnosis criteria. In particular modeling of the fault-free condition of nonlinear systems with the use of nonlinear regressors (e.g. neural networks) and the associated nonlinear least squares techniques are explained. Neural networks with Gauss-Hermite activation functions are introduced as an approach for modeling nonlinear dynamical systems, and particularly electric power generation and electric traction and propulsion systems. Besides, the statistical processing of the residuals of these neural networks, which are the differences between the outputs of the neural models and the outputs of the monitored system is shown to provide clear indications about the existence of faults.

In the present monograph on *Fault Diagnosis for Electric Power Systems and Electric Vehicles*, the following issues are analyzed: Chapter 1 is on Fault diagnosis with model-based and model-free techniques (Model-based fault diagnosis techniques and Model-free fault diagnosis techniques). Chapter 2 is on Control and fault diagnosis for Synchronous Generator-based renewable energy systems (Control of the marine-turbine and synchronous-generator unit and Fault diagnosis of the marine-turbine and synchronous-generator unit). Chapter 3 is on Fault diagnosis for electricity microgrids and gas processing units (Fault diagnosis for electric power DC microgrids and Fault diagnosis for electrically actuated gas compressors). Chapter 4 is on Fault diagnosis for gas and steam-turbine power generation units (Fault diagnosis for the gas-turbine and Synchronous Generator electric power unit and for the steam-turbine and synchronous generator power unit). Finally, Chapter 5 is on Fault diagnosis for wind power units and for the distribution grid (Fault diagnosis for wind power generators and Fault diagnosis for the electric power distribution grid).

Through the detailed and in depth treatment of the aforementioned topics, this monograph is expected to have a meaningful contribution to the members of the research, academic and engineering community. It is anticipated that the monograph will be particularly useful to researchers and university tutors working on fault diagnosis problems of electric power systems and of electric traction and propulsion systems.

Gerasimos Rigatos	Masoud Abbaszadeh	Mohamed Hamida	Pierluigi Siano
Athens, Greece	New York, USA	Nantes, France	Salerno, Italy
March, 2024	March, 2024	March, 2024	March, 2024

About the authors

Dr. Gerasimos Rigatos obtained his diploma (1995) and his Ph.D. (2000) both from the Department of Electrical and Computer Engineering, of the National Technical University of Athens (NTUA), Greece. In 2001 he was a post-doctoral researcher at IRISA-INRIA, Rennes, France. He is currently a Research Director (Researcher Grade A') at the Industrial Systems Institute, Greece. He is a Senior Member of IEEE, and a Member and CEng of IET. He has led several research cooperation agreements and projects which have given accredited results in the areas of nonlinear control, nonlinear filtering and control of distributed parameter systems. His results appear in 8 research monographs and in several journal articles. According to Elsevier Scopus his research comprising 135 journal articles where he is the first or sole author, has received more than 3000 citations with an H-index of 26. Since 2007, he has been awarded visiting professor positions at several academic institutions (University Paris XI, France, Harper-Adams University College, UK, University of Northumbria, UK, University of Salerno, Italy, Ecole Centrale de Nantes, France). He is an editor of the Journal of Information Sciences, the Journal of Advanced Robotic Systems and of the SAE Journal of Electrified Vehicles.

Dr. Masoud Abbaszadeh obtained a B.Sc and a M.Sc in Electrical Engineering from Amirkabir University of Technology and Sharif University of Technology, in Iran, respectively. Next, he received a Ph.D. degree in Electrical Engineering (Controls) in 2008 from the University of Alberta, Canada. From 2008 to 2011, he was with Maplesoft, Waterloo, Ontario, Canada, as a Research Engineer. He was the principal developer of MapleSim Control Design Toolbox and was a member of a research team working on the Maplesoft-Toyota joint projects. From 2011 to 2013, he was a Senior Research Engineer at United Technologies Research Center, East Hartford, CT, USA, working on advanced control systems, and complex systems modeling and simulation. Currently he is a Principal Research Engineer at GE Research Center, Niskayuna, NY, USA. He also holds an Adjunct Professor position at Rensselaer Polytechnic Institute, NY, USA. He has over 150 peer-reviewed papers, 9 book chapters, and holds 39 issued US patents, with over 40 more patents pending.. His research interests include estimation and detection theory, robust and nonlinear control, and machine learning with applications in diagnostics, cyber-physical resilience and autonomous systems. He serves as an Associate Editor of IEEE Transactions on Control Systems Technology, and a member of IEEE CSS Conference Editorial Board.

Dr. Mohamed-Assaad Hamida was born in El Oued, Algeria, in 1985. He received the B.Sc . degree in electrical engineering from the University of Batna, Batna, Algeria, in 2009, the M.Sc. degree in automatic control from Ecole Nationale Superieure d'Ingenieurs de Poitiers (ENSIP), Poitiers, France, in 2010, and the Ph.D degree in automatic control and electrical engineering from Ecole centrale de Nantes, Nantes, France, in 2013. From 2013 to 2017, he was an Associate Professor of Electrical Engineering with the University of Ouargla, Algeria. In 2017, he joined the Ecole Centrale de Nantes and the Laboratory of Digital Sciences of Nantes (LS2N), as an Associate Professor. Dr. Hamida is the local coordinator of the European project E-PiCo on Electric Vehicles Propulsion and Control at Ecole Centrale of Nantes and the head of the real-time systems unit in the same university. His research interests include robust nonlinear control (higher order sliding mode, backstepping, adaptive control, optimal control), theoretical aspects of nonlinear observer design, control and fault diagnosis of electrical systems and renewable energy applications. His current research interests include robust nonlinear control, theoretical aspects of nonlinear observer design, control, and fault diagnosis of electrical systems and renewable energy applications.

Dr. Pierluigi Siano received the M.Sc. degree in electronic engineering and the Ph.D. degree in information and electrical engineering from the University of Salerno, Salerno, Italy, in 2001 and 2006, respectively. He is Full Professor of Electrical Power Systems and Scientific Director of the Smart Grids and Smart Cities Laboratory with the Department of Management and Innovation Systems, University of Salerno. Since 2021 he has been a Distinguished Visiting Professor in the Department of Electrical and Electronic Engineering Science, University of Johannesburg. His research activities are centered on demand response, energy management, the integration of distributed energy resources in smart grids, electricity markets, and planning and management of power systems. In these research fields, he has co-authored more than 700 articles including more than 410 international journals that received in Scopus more than 19200 citations with an H-index equal to 66. Since 2019 he has been awarded as a Highly Cited Researcher in Engineering by Web of Science Group. He has been the Chair of the IES TC on Smart Grids. He is Editor for the Power & Energy Society Section of IEEE Access, IEEE Transactions on Power Systems, IEEE Transactions on Industrial Informatics, IEEE Transactions on Industrial Electronics, and IEEE Systems.

1 Fault diagnosis with model-based and model-free techniques

1.1 MODEL-BASED FAULT DIAGNOSIS

1.1.1 OUTLINE

The use of fault detection and isolation methods is necessary for deterring critical conditions and for enabling reliable functioning and preventive maintenance in electric power systems, as well as in electric traction and propulsion systems. When the dynamics of the monitored system is explicitly known, one can develop model-based fault diagnosis techniques which rely on the statistical processing of the differences between the outputs of the system and the outputs of the model, which represents the system in fault-free mode. A related example will be given in the case of three-phase voltage source inverters. Voltage source inverters are critical components of electric power conversion systems. Actually, the use of three-phase inverters is widespread, in both actuation and traction systems and in electric power distribution systems. A primary field of application of three-phase inverters (or converters) is the control of electric motors (such as induction motors and permanent magnet synchronous motors). Consequently, inverters are a prerequisite for motion generation and for actuation in several industrial tasks [38], [254]. Besides, inverters are of major importance for the traction system of electric vehicles and trains [283], [47]. Additionally, multi-phase inverters are essential for the control of multi-phase synchronous motors, as for instance in the traction of heavy-duty vehicles and in the propulsion of electric ships. In parallel, a large application field for inverters is that of renewable energy. Three-phase inverters enable to connect and synchronize DC power units with the main electricity grid [39], [40], [113], [222]. The faultless functioning of inverters is critical for maintaining the quality of the grid's electric power, minimizing disturbances in the electricity network, and reducing also power losses in DC to AC power conversion. Finally, inverters find use as active power filters, aiming at generating harmonics that eliminate disturbances in the grid's voltage.

Due to being exposed to harsh operating conditions, three-phase inverters undergo failures [264], [118], [93], [217]. Furthermore, when inverters function as part of networked control schemes, they become exposed to cyber-attacks targeting their control and data acquisition software. It is important to accomplish early fault detection and incipient failure diagnosis for inverters so as to avoid excessive damage of this equipment and to take action for its repair. Major objectives in fault detection and isolation for three-phase inverters, as well as in the diagnosis of cyber-attacks against the inverters' control software, are (i) the early diagnosis of malfunctioning [209], [68], [153], [125], (ii) the spotting of incipient parametric changes [222], [7], [181] and (iii) the development of systematic methods for the definition of fault thresholds [75], [36], [278], [4]. Fault diagnosis for the nonlinear dynamic model of three-phase inverters is a non-trivial problem that has been treated in several cases with the use of state observers and statistical filters [195], [119], [110], [276], [188]. Additional results on Kalman Filtering, for state and parameters estimation and for fault diagnosis purposes can be found in [291], [292], [158], [157], [221]. Moreover, results on adaptive estimators and adaptive Kalman Filtering, which can be used in fault diagnosis applications, can be found in [271], [273], [274], [275], [284]. In the present chapter, a systematic method is developed for fault diagnosis in three-phase inverters, as well as for the detection of cyber-attacks and malicious signals affecting their control loop. The method relies on the state vector and outputs estimation for the inverter with the use of a robust filtering scheme, known as H-infinity Kalman Filter. Since the H-infinity Kalman Filter is primarily addressed to linear dynamical systems, to enable its use in the nonlinear model of the inverter, the state-space description of the inverter undergoes linearization through Taylor series expansion. The linearization is performed around a temporary operating point, which is recomputed at each time step of the condition monitoring method.

Besides, the proposed condition monitoring method defines fault thresholds in a systematic and precise manner. The H-infinity Kalman Filter is considered to emulate the function of the inverter in fault-free cases. The filter's output is compared against the real output of the inverter, thus generating the residual vectors' sequence. It is shown that the sum of the square of the residuals vectors, weighted with their inverse covariance matrix stands for a stochastic variable which follows the χ^2 distribution [198], [18], [189]. Next, by using the confidence intervals of this distribution one can conclude with a high confidence level (for instance a certainty measure of the order of 95% or 98%) the normal functioning of the system or the appearance of a fault. As long as this stochastic variable falls between the upper and lower bounds of the confidence interval, one can infer that the inverter is in the fault-free mode. On the other side, when the upper or lower bound is persistently exceeded, then the existence of a fault can be diagnosed and an alarm can be launched. Furthermore, by applying the statistical test selectively to subspaces of the state-space description of the inverter, one can also achieve fault isolation.

1.1.2 DYNAMIC MODEL OF THE THREE-PHASE INVERTER

1.1.2.1 Dynamics of the inverter

The inverter's (DC to AC converter's) circuit is depicted in Fig. 1.1. By applying Kirchoff's voltage and current laws, one obtains [198]

$$\frac{d}{dt}i_I = \frac{1}{L_f}V_I - \frac{1}{L_f}V_L$$
$$\frac{d}{dt}V_L = \frac{1}{C_f}i_I - \frac{1}{C_f}i_L \tag{1.1}$$

For the representation of the voltage and current variables, denoted as $X = \{I, V\}$ in the ab static reference frame, one has

$$X_{ab} = X_a e^{j0} + X_b e^{\frac{j2\pi}{3}} + X_c e^{\frac{j4\pi}{3}} \tag{1.2}$$

which finally gives a complex variable of the form

$$X_{ab} = X_a + jX_b \tag{1.3}$$

Next, the voltage and current variables are represented in the rotating dq reference frame [198]. It holds that $X_{dq} = x_d + jx_q$ and

$$X_{dq} = X_{ab}e^{-j\theta} \Rightarrow X_{ab} = X_{dq}e^{j\theta}$$
$$\text{where } \theta(t) = \int_0^t \omega(t)dt + \theta_0 \tag{1.4}$$

By differentiating with respect to time, one obtains the following description of the system's dynamics

$$\dot{X}_{ab} = \frac{d}{dt}X_{dq} + j\omega X_{dq} \tag{1.5}$$

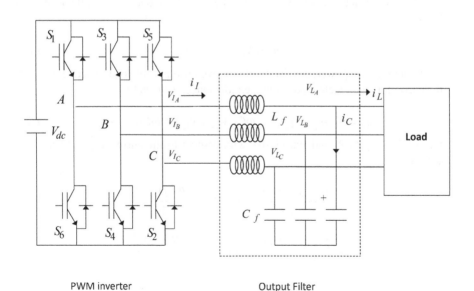

Figure 1.1   Circuit of the three-phase voltage inverter

Thus, for the current and voltage variables one has respectively,

$$\begin{aligned}
\dot{i}_{I,ab} &= \tfrac{d}{dt} i_{I,dq} + (j\omega) i_{I,dq} \\
\dot{V}_{L,ab} &= \tfrac{d}{dt} V_{L,dq} + (j\omega) V_{L,dq}
\end{aligned} \tag{1.6}$$

By substituting Eq. (1.6) into Eq. (1.1), one obtains

$$\begin{aligned}
\tfrac{d}{dt} i_{I,dq} + j\omega i_{I,dq} &= \tfrac{1}{L_f} V_{I,dq} - \tfrac{1}{L_f} V_{L,dq} \\
\tfrac{d}{dt} V_{L,dq} + j\omega V_{L,dq} &= \tfrac{1}{C_f} i_{I,dq} - \tfrac{1}{C_f} i_{L,dq}
\end{aligned} \tag{1.7}$$

Using Eq. (1.7) and by rearranging rows, one finally obtains the inverter's dynamic model expressed in the dq reference frame:

$$\begin{aligned}
\tfrac{d}{dt} V_{L,d} &= \omega V_{L,q} + \tfrac{1}{C_f} i_{I,d} - \tfrac{1}{C_f} i_{L,d} \\
\tfrac{d}{dt} V_{L,q} &= -\omega V_{L,d} + \tfrac{1}{C_f} i_{I,q} - \tfrac{1}{C_f} i_{L,q} \\
\tfrac{d}{dt} i_{I,d} &= \omega i_{I,q} + \tfrac{1}{L_f} V_{I,d} - \tfrac{1}{L_f} V_{L,d} \\
\tfrac{d}{dt} i_{I,q} &= -\omega i_{I,d} + \tfrac{1}{L_f} V_{I,q} - \tfrac{1}{L_f} V_{L,q}
\end{aligned} \tag{1.8}$$

The state vector of the system is taken to be $\tilde{X} = [V_{L_d}, V_{L_q}, i_{I,d}, i_{I,q}]^T$, while the control input is taken to be the vector $\tilde{U} = [V_{I,d}, V_{I,q}]$. The load currents $i_{L,d}$ and $i_{L,q}$ are taken to be unknown parameters, which can be considered as perturbation terms. Alternatively, these currents can be expressed as functions of the inverter's active and reactive power. In the latter approach, one has that the active power of the inverter is [118]

$$p_f = V_{L_d} i_{L_d} + V_{L_q} i_{L_q} \tag{1.9}$$

while the reactive power, consisting of reactive power at the load, reactive power at the capacitor, and reactive power at the inductance, is given by

$$q_f = V_{L_q} i_{L_d} - V_{L_d} i_{L_q} - \omega C_f (V_{L_d}^2 + V_{L_q}^2) + \omega L_f (i_{I,d}^2 + i_{I,q}^2) \tag{1.10}$$

By solving Eq. (1.9) and Eq. (1.10), with respect to the currents i_{L_d} and i_{L_q}, one obtains

$$i_{L_d} = \frac{p_f V_{L_d} + q_f V_{L_q}}{V_{L_d}^2 + V_{L_q}^2} + \omega C_f V_{L_q} - \frac{\omega L_f V_{L_q}(i_{I_d}^2 + i_{I_q}^2)}{(V_{L_d}^2 + V_{L_q}^2)} \tag{1.11}$$

$$i_{L_q} = \frac{p_f V_{L_q} - q_f V_{L_d}}{V_{L_d}^2 + V_{L_q}^2} - \omega C_f V_{L_d} + \frac{\omega L_f V_{L_d}(i_{I_d}^2 + i_{I_q}^2)}{(V_{L_d}^2 + V_{L_q}^2)} \tag{1.12}$$

Using Eq. (1.8) and Eq. (1.11), Eq. (1.12), one obtains the state-space description of the inverter's dynamics

$$
\frac{d}{dt}\begin{pmatrix} V_{L_d} \\ V_{L_q} \\ i_{L_d} \\ i_{L_q} \end{pmatrix} = \begin{pmatrix} \omega V_{L_q} + \frac{1}{C_f} i_{L_d} - \frac{1}{C_f}\left[\frac{p_f V_{L_d} + q_f V_{L_q}}{V_{L_d}^2 + V_{L_q}^2}\right] + \omega C_f V_{L_q} - \frac{\omega L_f V_{L_q}(i_{L_d}^2 + i_{L_q}^2)}{(V_{L_d}^2 + V_{L_q}^2)} \\ -\omega V_{L_d} + \frac{1}{C_f} i_{L_q} - \frac{1}{C_f}\left[\frac{p_f V_{L_q} - q_f V_{L_d}}{V_{L_d}^2 + V_{L_q}^2}\right] - \omega C_f V_{L_d} + \frac{\omega L_f V_{L_d}(i_{L_d}^2 + i_{L_q}^2)}{(V_{L_d}^2 + V_{L_q}^2)} \\ \omega i_{L_q} - \frac{1}{L_f} V_{L_d} \\ -\omega i_{L_d} - \frac{1}{L_f} V_{L_q} \end{pmatrix} +
$$

$$
+ \begin{pmatrix} 0 & 0 \\ 0 & 0 \\ \frac{1}{L_f} & 0 \\ 0 & \frac{1}{L_f} \end{pmatrix} \begin{pmatrix} V_{I_d} \\ V_{I_q} \end{pmatrix} \tag{1.13}
$$

while the measurement equation of the inverter's model is

$$
\begin{pmatrix} y_1 \\ y_2 \end{pmatrix} = \begin{pmatrix} V_{L_d} \\ V_{L_q} \end{pmatrix} = \begin{pmatrix} 1 & 0 & 0 & 0 \\ 0 & 1 & 0 & 0 \end{pmatrix} \begin{pmatrix} V_{L_d} \\ V_{L_q} \\ i_{L_d} \\ i_{L_q} \end{pmatrix} \tag{1.14}
$$

and by using the state variables notation $x_1 = V_{L_d}$, $x_2 = V_{L_q}$, $x_3 = i_{L_d}$ and $x_4 = i_{L_q}$, one has

$$
\frac{d}{dt}\begin{pmatrix} x_1 \\ x_2 \\ x_3 \\ x_4 \end{pmatrix} = \begin{pmatrix} \omega x_2 + \frac{1}{C_f} x_3 - \frac{1}{C_f}\left[\frac{p_f x_1 + q_f x_2}{x_1^2 + x_2^2}\right] + \omega C_f x_2 - \frac{\omega L_f x_2(x_3^2 + x_4^2)}{(x_1^2 + x_2^2)} \\ -\omega x_1 + \frac{1}{C_f} x_4 - \frac{1}{C_f}\left[\frac{p_f x_2 - q_f x_1}{x_1^2 + x_2^2}\right] - \omega C_f x_1 + \frac{\omega L_f x_1(x_3^2 + x_4^2)}{(x_1^2 + x_2^2)} \\ \omega x_4 - \frac{1}{L_f} x_1 \\ -\omega x_3 - \frac{1}{L_f} x_2 \end{pmatrix} +
$$

$$
+ \begin{pmatrix} 0 & 0 \\ 0 & 0 \\ \frac{1}{L_f} & 0 \\ 0 & \frac{1}{L_f} \end{pmatrix} \begin{pmatrix} u_1 \\ u_2 \end{pmatrix} \tag{1.15}
$$

while the measurement equation of the inverter's model is

$$
\begin{pmatrix} y_1 \\ y_2 \end{pmatrix} = \begin{pmatrix} V_{L_d} \\ V_{L_q} \end{pmatrix} = \begin{pmatrix} 1 & 0 & 0 & 0 \\ 0 & 1 & 0 & 0 \end{pmatrix} \begin{pmatrix} x_1 \\ x_2 \\ x_3 \\ x_4 \end{pmatrix} \tag{1.16}
$$

thus, the inverter's model is written in the nonlinear state-space form

$$\dot{x} = f(x) + G(x)u$$
$$y = h(x)$$

(1.17)

where $f(x) \in R^{4 \times 1}$, $G(x) \in R^{4 \times 1}$ and $h(x) \in R^{2 \times 4}$.

1.1.2.2 Faults affecting the three-phase inverter

The Kalman Filter's outputs are compared at each sampling instance against the outputs measured from the three-phase inverter, thus providing the residuals' sequence. Indicative faults and cyberattacks or malignant human intrusions that may affect the three-phase inverter, which comprises the H-bridge (transistors) circuit and the LC circuit at its output, are shown in Fig. 1.2. The three-phase voltage source inverter consists of (a) a set of transistors that form the H-bridge circuit and (b) an LC filter which is connected to the output of the H-bridge circuit. Faults at the H-bridge circuit take the form of additive disturbances at the voltage input variables V_{I_d}, V_{I_q}, or have the form of additive disturbances that affect the voltage variables V_{L_d}, V_{L_q} of the state-space model of the three-phase voltage source inverter, which is given in Eq. (1.13) and Eq. (1.14). Faults at the LC filter take the form of parametric changes which affect the coefficients L_f, C_f and ω that appear in the state-space model of Eq. (1.13) and (1.14). There are also faults that may affect the sensors that measure the input and output voltages of the inverter, as well as faults affecting the inputs and outputs of the Kalman Filter and the associated state estimation process.

With reference to Fig. 1.2, the faults of the three-phase inverter have been classified as follows: (i) Faults at the control input gains matrix of the inverter Δig_1, Δig_2, (ii) Faults at the drift vector $f(x)$ of the inverter's state-space model, (iii) Faults at the control inputs gain matrix $g(x)$ of the inverter's state-space model, (iv) Faults (additive disturbances) at the control inputs at the inverter Δu_1, Δu_2, (v) Faults at the inverter's control input sensors Δis_1, Δis_2, (vi) Faults at the inverters' output measurement sensors Δos_1, Δos_2 and (vii) Faults at the output of the Kalman Filter.

1.1.3 APPROXIMATE LINEARIZATION OF THE THREE-PHASE INVERTER

The dynamic model of the three-phase inverter has been shown to be

$$\dot{x} = f(x) + g(x)u$$

(1.18)

where the state vector is $x = [x_1, x_2, x_3, x_4]^T$, and vector fields $f(x)$, $g(x)$, which can be obtained from Eq. (1.15).

The three-phase inverter undergoes approximate linearization around the time-varying operating point (x^*, u^*), where x^* denotes the present value of the state vector

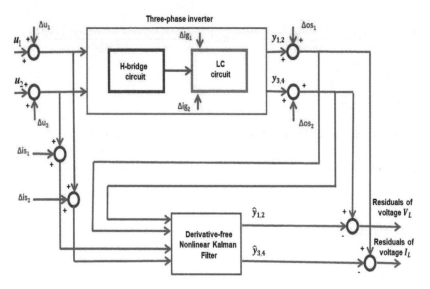

Figure 1.2   Faults and cyberattacks affecting the three-phase inverter

and u^* denotes the last sampled value of the control inputs vector. This linearization procedure requires the computation of Jacobian matrices and gives

$$\dot{x} = Ax + Bu + \tilde{d} \tag{1.19}$$

where $\tilde{d}$ is the modeling error due to approximate linearization and cut-off of higher order terms in the Taylor series expansion, while matrices A and B are obtained from the computation of the Jacobian matrices

$$A = \nabla_x[f(x) + g(x)u] \,|_{(x^*,u^*)} \tag{1.20}$$

or using that $g(x)$ does not contain elements depending of x, one has the following description for the Jacobian matrix A

$$A = \begin{pmatrix} \frac{\partial f_1}{\partial x_1} & \frac{\partial f_1}{\partial x_2} & \frac{\partial f_1}{\partial x_3} & \frac{\partial f_1}{\partial x_4} \\ \frac{\partial f_2}{\partial x_1} & \frac{\partial f_2}{\partial x_2} & \frac{\partial f_2}{\partial x_3} & \frac{\partial f_2}{\partial x_4} \\ \frac{\partial f_3}{\partial x_1} & \frac{\partial f_3}{\partial x_2} & \frac{\partial f_3}{\partial x_3} & \frac{\partial f_3}{\partial x_4} \\ \frac{\partial f_4}{\partial x_1} & \frac{\partial f_4}{\partial x_2} & \frac{\partial f_4}{\partial x_3} & \frac{\partial f_4}{\partial x_4} \end{pmatrix} \tag{1.21}$$

For the first row of the Jacobian matrix $A = \nabla_x(f(x))$, it holds that

$$\frac{\partial f_1}{\partial x_1} = -\frac{1}{C_f}\left[\frac{p_f(x_1^2+x_2^2)-(p_f x_1+q_f x_2)2x_1}{(x_1^2+x_2^2)^2} + \frac{\omega L_f x_2(x_3^2+x_4^2)2x_1}{(x_1^2+x_2^2)^2}\right],$$

$$\frac{\partial f_1}{\partial x_2} = -\frac{1}{C_f}\left[\frac{q_f(x_1^2+x_2^2)-(p_f x_1+q_f x_2)2x_2}{(x_1^2+x_2^2)^2}\right] + \omega C_f - \frac{\omega L_f(x_3^2+x_4^2)(x_1^2+x_2^2)-\omega L_f x_2(x_3^2+x_4^2)2x_2}{(x_1^2+x_2^2)^2}\right],$$

$$\frac{\partial f_1}{\partial x_3} = \frac{1}{C_f} - \frac{\omega L_f x_2 2x_3}{x_1^2+x_2^2},$$

$$\frac{\partial f_1}{\partial x_4} = -\frac{\omega L_f x_2 2x_4}{x_1^2+x_2^2}$$

For the second row of the Jacobian matrix $\nabla_x(f(x))$, it holds that

$$\frac{\partial f_2}{\partial x_1} = -\omega - \frac{1}{C_f}\left[\frac{-q_f(x_1^2+x_2^2)-(p_f x_2-q_f x_1)2x_1}{(x_1^2+x_2^2)^2}\right] - \omega C_f + \frac{\omega L_f(x_3^2+x_4^2)(x_1^2+x_2^2)-\omega L_f x_1(x_3^2+x_4^2)2x_1}{(x_1^2+x_2^2)^2}\right],$$

$$\frac{\partial f_2}{\partial x_2} = -\frac{1}{C_f}\left[\frac{p_f(x_1^2+x_2^2)-(p_f x_2-q_f x_1)2x_2}{(x_1^2+x_2^2)^2}\right] - \omega C_f - \frac{\omega L_f x_1(x_3^2+x_4^2)2x_2}{(x_1^2+x_2^2)^2}\right]$$

$$\frac{\partial f_2}{\partial x_3} = \frac{\omega L_f x_1 2x_3}{x_1^2+x_2^2}$$

$$\frac{\partial f_2}{\partial x_4} = \frac{1}{C_f} + \frac{\omega L_f x_1 2x_4}{x_1^2+x_2^2}$$

For the third row of the Jacobian matrix $\nabla_x(f(x))$, it holds that $\frac{\partial f_3}{\partial x_1} = -\frac{1}{L_f}, \frac{\partial f_3}{\partial x_2} = 0,$ $\frac{\partial f_3}{\partial x_3} = 0, \frac{\partial f_3}{\partial x_4} = \omega$

For the fourth row of the Jacobian matrix $\nabla_x(f(x))$, it holds that $\frac{\partial f_4}{\partial x_1} = 0, \frac{\partial f_4}{\partial x_2} = -\frac{1}{L_f},$ $\frac{\partial f_4}{\partial x_3} = -\omega, \frac{\partial f_4}{\partial x_4} = 0$

Moreover, it holds that

$$B = \nabla_u[f(x)+g(x)u]\,|_{(x^*,u^*)} \Rightarrow$$

$$B = g(x) \Rightarrow B = \begin{pmatrix} 0 & 0 \\ 0 & 0 \\ \frac{1}{L_f} & 0 \\ 0 & \frac{1}{L_f} \end{pmatrix} \tag{1.22}$$

Thus, after linearization around its current operating point, the model of the three-phase inverter is written in the form of Eq (1.19), that is $\dot{x} = Ax + Bu + \tilde{d}$. For the approximately linearized model of the inverter an H-infinity controller has been developed. The controller has the form

$$u(t) = -Ke(t) \tag{1.23}$$

with $K = \frac{1}{r}B^T P$ where P is a positive definite symmetric matrix, which is obtained from the solution of the Riccati equation and $e = x - x_d$ to be the state vector's tracking error

$$A^T P + PA + Q - P(\frac{2}{r}BB^T - \frac{1}{\rho^2}LL^T)P = 0 \tag{1.24}$$

while Q is a positive semi-definite symmetric matrix.

1.1.4 STATISTICAL FAULT DIAGNOSIS WITH THE H-INFINITY KALMAN FILTER

1.1.4.1 The H-infinity Kalman Filter

The chapter's H-infinity Kalman filtering approach for state estimation and fault-free modeling of the three-phase inverter is conceptually simple and has clear implementation stages [198], [195]. The nonlinear state-space model of the monitored system undergoes approximate linearization around a temporary operating point, which is recomputed at each time step of the estimation algorithm. For the approximately linearized model, state estimation is performed with the use of the H-infinity Kalman Filter. The effects of measurement noise, as well as the modeling error which is due to truncation of higher-order terms in the Taylor series expansion is considered to be a perturbation, which is asymptotically compensated by the robustness of the H-infinity Kalman Filter.

The H-infinity Kalman Filter is taken to represent the fault-free functioning of the power unit. The optimization conditions that result in the recursive computation of the H-infinity Kalman Filter are given in Chapter 1. The recursion of the H-infinity Kalman Filter can be formulated again in terms of a *measurement update* and a *time update* part [78], [226]

Measurement update:

$$D(k) = [I - \theta W(k)P^-(k) + C^T(k)R(k)^{-1}C(k)P^-(k)]^{-1}$$
$$K(k) = P^-(k)D(k)C^T(k)R(k)^{-1} \tag{1.25}$$
$$\hat{x}(k) = \hat{x}^-(k) + K(k)[y(k) - C\hat{x}^-(k)]$$

Time update:

$$\hat{x}^-(k+1) = A(k)x(k) + B(k)u(k)$$
$$P^-(k+1) = A(k)P^-(k)D(k)A^T(k) + Q(k) \tag{1.26}$$

where it is assumed that the parameter θ is sufficiently small to maintain

$$P^-(k)^{-1} - \theta W(k) + C^T(k)R(k)^{-1}C(k) \tag{1.27}$$

positive definite. When $\theta = 0$, the H_∞ Kalman Filter becomes equivalent to the standard Kalman Filter. It is noted that apart from the process noise covariance matrix $Q(k)$ and the measurement noise covariance matrix $R(k)$, the H_∞ Kalman filter requires tuning of the weight matrices L and S, as well as of parameter θ.

To elaborate on the matrices which appear in the *Measurement update* part and in the *Time update part* of the H-infinity Kalman Filter, the following can be noted: Matrix $R(k)$ is the measurement noise covariance matrix, that is the covariance matrix of the measurement error vector of the system. Matrix $P^-(k)$ is the a-priori state vector estimation error covariance matrix of the system, that is the covariance matrix of the state vector estimation error prior to receiving the updated measurement of the system's outputs. Matrix $W(k)$ is a weight matrix which defines the significance to be attributed by the H-infinity Kalman Filter in the minimization of the state vector's estimation error, relatively to the effects that the noise affecting the system may have. Finally, matrix $D(k)$ stands for a modified a-posteriori state vector estimation error covariance matrix, that is the covariance matrix of the state vector estimation error after receiving the updated measurement of the system's outputs. Conclusively, the H-infinity Kalman Filter retains the structure of the typical Kalman Filter, that is a recursion in discrete time comprising a Time update part (computation of variables prior to receiving measurements) and a Measurement update part (computation of variables after measurements have been received). There is a modified a-posteriori state vector estimation error covariance matrix, which in turn takes into account a weight matrix that defines the accuracy of the state estimation under the effects of elevated noise. The stability of the H-infinity Kalman Filter and the convergence of the state estimation procedure that is performed by it relies on the detectability/observability properties of the monitored system and not on initialization of the matrices that appear in the filter's recursion.

1.1.4.2   Fault detection

The residuals' sequence, that is the differences between (i) the real outputs of the three-phase inverter and (ii) the outputs estimated by the Kalman Filter (Fig. 2.37) is a discrete error process e_k with dimension $m \times 1$ (here $m = N$ is the dimension of the output measurements vector). Actually, it is a zero-mean Gaussian white-noise process with covariance given by E_k.

A conclusion can be stated based on a measure of certainty that the three-phase inverter has neither been subjected to a fault nor to a cyberattack. To this end, the following *normalized error square* (NES) is defined [195]

$$\varepsilon_k = e_k^T E_k^{-1} e_k \qquad (1.28)$$

The normalized error square follows a χ^2 distribution. An appropriate test for the normalized error sum is to numerically show that the following condition is met within a level of confidence (according to the properties of the χ^2 distribution)

$$E\{\varepsilon_k\} = m \qquad (1.29)$$

This can be achieved using statistical hypothesis testing, which is associated with confidence intervals. A 95% confidence interval is frequently applied, which is specified using the probability region $100(1 - a)$ with $a = 0.05$. Actually, a two-sided probability region is considered cutting-off two end tails of 2.5% each. For M runs the normalized error square that is obtained is given by

$$\bar{\varepsilon}_k = \frac{1}{M} \sum_{i=1}^{M} \varepsilon_k(i) = \frac{1}{M} \sum_{i=1}^{M} e_k^T(i) E_k^{-1}(i) e_k(i) \qquad (1.30)$$

where ε_i stands for the i-th run at time t_k. Then $M\bar{\varepsilon}_k$ will follow a χ^2 density with Mm degrees of freedom. This condition can be checked using a χ^2 test. The hypothesis holds, if the following condition is satisfied

$$\bar{\varepsilon}_k \in [\zeta_1, \zeta_2] \tag{1.31}$$

where ζ_1 and ζ_2 are derived from the tail probabilities of the χ^2 density. For example, for $m = 20$ (dimension of the measurements vector) and $M = 100$ (total number of the output vector's samples) one has $\chi^2_{Mm}(0.025) = 1878$ and $\chi^2_{Mm}(0.975) = 2126$. Using that $M = 100$, one obtains $\zeta_1 = \chi^2_{Mm}(0.025)/M = 18.78$ and $\zeta_2 = \chi^2_{Mm}(0.975)/M = 21.26$.

1.1.4.3 Fault isolation

By applying the statistical test into the individual components of the three-phase inverter, it is also possible to find out the specific component that has been subjected to a fault or cyberattack [195], [198]. For an inverter of n parameters suspected for change one has to carry out n χ^2 statistical change detection tests, where each test is applied to the subset that comprises parameters $i-1$, i and $i+1$, $i = 1, 2, \cdots, n$.

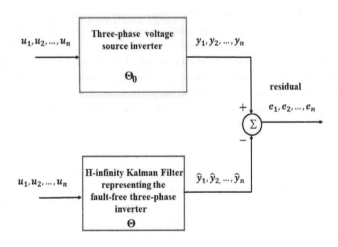

Figure 1.3 Residuals'generation for the three-phase inverter, with the use of Kalman Filtering

Actually, out of the n χ^2 statistical change detection tests, the ones that exhibit the highest score are those that identify the parameter that has been subjected to change.

In the case of multiple faults, one can identify the subset of parameters that has been subjected to change by applying the χ^2 statistical change detection test according to a combinatorial sequence. This means that

$$\binom{n}{k} = \frac{n!}{k!(n-k)!} \tag{1.32}$$

tests have to take place, for all clusters in the inverter, that finally comprise n, $n-1$, $n-2$, $\cdots$, 2, 1 parameters. Again the χ^2 tests that give the highest scores indicate the parameters which are most likely to have been subjected to change.

1.1.5 SIMULATION TESTS

The performance of the proposed fault diagnosis method for three-phase inverters has been tested through simulation experiments. The obtained results are depicted in Fig. 1.4 to Fig. 1.9. The capability of the method for detecting incipient fault and small parametric changes has been confirmed. Actually, the following test cases have been presented: (i) functioning of the three-phase inverter in the fault-free state, as shown in Fig. 1.4; (ii) functioning of the three-phase inverter under parametric change (fault) affecting the drift vector f in its state-space description, as shown in Fig. 1.5; (iii) functioning of the three-phase inverter under parametric change (fault) affecting the control inputs gain matrix g of its state-space description, as shown in Fig. 1.6; (iv) functioning of the three-phase inverter under additive perturbation Δu affecting its control inputs, as shown in Fig. 1.7; (v) functioning of the three-phase inverter under additive perturbations affecting the control inputs of the Kalman Filter that is used for condition monitoring purposes as shown in Fig. 1.8 and (vi) functioning of the three-phase inverter under a disturbance that affects the sensors measuring its outputs, as shown in Fig. 1.9.

To perform fault detection and isolation for the three-phase inverter the statistical properties of the residuals of the H-infinity Kalman Filter have been used. The dimension of the measurements vector was equal to $n = 4$, and this was also the dimension of the residuals' vector. This signifies that in the fault-free case the considered statistical test should have a value that is very close to the mean value of the χ^2 distribution with $n = 4$ degrees of freedom. Consequently, in the fault-free functioning of the inverter the statistical test should have a value equal to 4. The 98% confidence intervals about finding the system in the fault-free case were determined by the lower and upper bounds $L = 3.89$ and $U = 4.11$ respectively. As confirmed by the simulation experiments, as long as the value of the statistical test falls within these confidence intervals, it is concluded that the functioning of the inverter is normal. On the other side, when the above noted upper or lower bounds are exceeded,

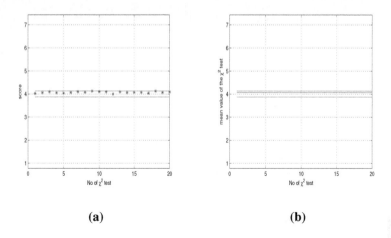

(a)   **(b)**

Figure 1.4   (a) Result of successive χ^2 tests relying on the H-infinity Kalman Filter in case of no-fault at the inverter and (b) mean value of the χ^2 tests relying on the H-infinity Kalman Filter in case of no-fault at the inverter

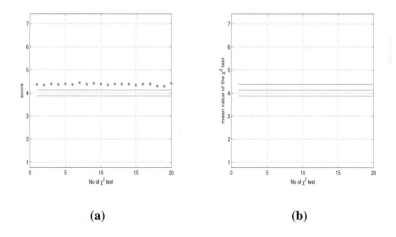

(a)   **(b)**

Figure 1.5   (a) Result of successive χ^2 tests relying on the H-infinity Kalman Filter in case of parametric change at the inverter's input gains matrix G and (b) mean value of the χ^2 tests relying on the H-infinity Kalman Filter in case of parametric change at the inverter's input gains matrix G

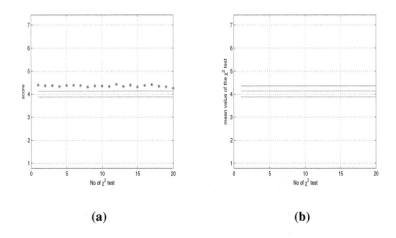

(a)   (b)

Figure 1.6 (a) Result of successive χ^2 tests relying on the H-infinity Kalman Filter in case of parametric change at the inverter's drift vector F and (b) mean value of the χ^2 tests relying on the H-infinity Kalman Filter in case of parametric change at the inverter's drift vector F

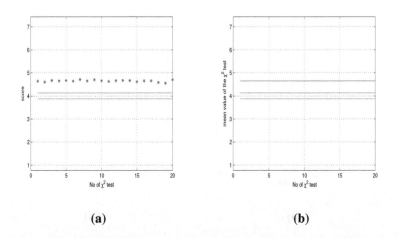

(a)   (b)

Figure 1.7 (a) Result of successive χ^2 tests relying on the H-infinity Kalman Filter in case of additive disturbance at the inverter's control inputs u and (b) mean value of the χ^2 tests relying on the H-infinity Kalman Filter in case of additive disturbance at the inverter's control input u

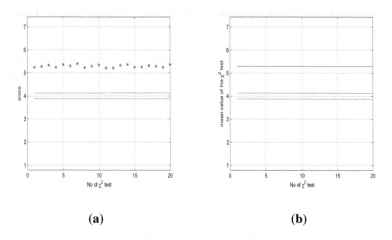

(a) **(b)**

Figure 1.8 (a) Result of successive χ^2 tests relying on the HKF in case of additive disturbance at the inputs of the H-infinity Kalman Filter and (b) mean value of the χ^2 tests relying on the H-infinity Kalman Filter in case of additive disturbance at the inputs of the HKF

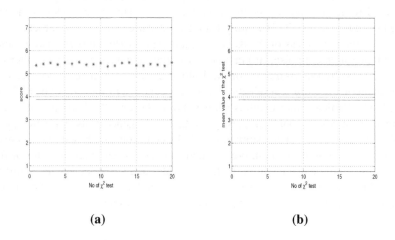

(a) **(b)**

Figure 1.9 (a) Result of successive χ^2 tests relying on the H-infinity Kalman Filter in case of disturbance at the sensors measuring the inverter's outputs and (b) mean value of the χ^2 tests relying on the H-infinity Kalman Filter in case of disturbance at the sensors measuring the inverter's outputs

it can be concluded that the system has undergone a failure and an alarm can be launched.

There can be comparison of the proposed nonlinear H-infinity Kalman Filter used for state estimation in the dynamic model of the three-phase voltage against (a) Luenberger-type state observers applied on the linearized model of the system, (b) sliding-mode observers, (c) the Extended Kalman Filter, (d) the Sigma-Point Kalman Filter and the Particle Filter and (e) diffeomorphisms-based Kalman Filter estimators. In case of (a), Luenberger-type state-observers do not take into account measurement noise in their estimation process, therefore from the point of view statistical estimation theory they are incomplete. Besides, the computation of the gains of these observers has to take place through a pole-placement technique, which is to a large extent empirical and does not take into account the optimality conditions of the Kalman Filter. In case of (b), the sliding-mode observer exhibits the previously noted flaws, that is the statistics of the measurement noise are not considered at all by the estimation process. Besides, the selection of the observer's feedback gains is based on crude stability conditions and does not take into account the fine optimality criteria which are met in Kalman Filter. Moreover, the existence of switching control terms in the observer's gains results into unnecessary jittering of the estimated values and the convergence of estimated variables to the real variables may not be sufficiently smooth. In case of (c), the Extended Kalman Filter has inferior performance comparing to the nonlinear H-infinity Kalman Filter. Actually, EKF uses the standard Kalman Filter recursion on the linearized state-space model of the monitored system whereas the H-infinity Kalman Filter uses an optimized recursion that improves the robustness of the Kalman Filter against external perturbations and elevated measurement noise, (iv) in case of (d), the Sigma Point Kalman Filter is not characterized by the optimality of the Kalman Filter while its convergence is also dependant on the empirical selection of several parameters and cannot be always ensured. The Particle Filter exhibits similar flaws. Besides, the Particle Filter suffers from a high computational complexity and the convergence of its estimation process is slow. In case of (e). state estimation with diffeomorphisms-based filters considers the transformation of the monitored system's dynamics into equivalent linear forms after using complicated changes of state variables and subsequently the application of the Kalman Filter on the linearized equivalent model of the system. Such transformations, coming often from Lie algebra may be too complex.

1.2 MODEL-FREE FAULT DIAGNOSIS

1.2.1 OUTLINE

When the dynamics of the monitored system is not explicitly known, then a neural network can be used to develop a model of the system through the processing of input-output raw data. This approach enables model-free fault diagnosis and a related example will be shown in the case of a Permanent Magnet Synchronous Linear

Motor. PMLSMs, that is Permanent Magnet Linear Synchronous Machines, are widely used in industry in tasks where linear motion has to be generated, as well as in railway transportation for instance in the propulsion of trains. Besides, they are used for electric power generation, for instance in the case of wave energy conversion [198], [134], [200], [85]. In particular Permanent Magnet Linear Synchronous Machines are considered to be obtained from the rotary Permanent Magnet Synchronous Machines after unfolding and arranging sequentially both the stator's windings and the array of permanent magnets of the rotor [117], [31], [151], [149]. PMLSMs comprise a primary (moving) part on which usually the permanent magnets are placed, as well as a secondary (stationary) part where the three-phase windings are placed. Due to the magnetic field that is developed at the secondary part and its effect on the permanent magnets of the primary part, a propulsion force is exerted on the primary component of the motor making it move. The dynamic model of the PMLSM is a nonlinear multi-variable one the related state estimation and condition monitoring problem is a non-trivial task [108], [255], [164], [220], [210].

Due to harsh operating conditions PMLSMs often undergo failures. Early detection of incipient faults can deter the occurrence of irreversible damage in such machines as well as the appearance of critical conditions, such as the interruption of the functioning of industrial production units, of transportation means or of power generation units [18], [19], [269]. The main approaches for solving the related fault detection and isolation problems can be classified in time-domain methods and in frequency domain methods [189], [190], [104]. One can also classify the fault diagnosis methods for PMLSMs into model-based and model-free ones. In the model-based methods, it is assumed that the dynamic model of the electric machines and its parameters in the fault-free condition are previously known [268], [26], [270]. In the model-free methods, it is considered that both the dynamic model of the PMLSMs and its parameters in the fault-free condition are unknown and have to be estimated through the processing of input-output data [96], [213], [165], [48], [110], [172]. Often the fault threshold in the condition monitoring of PMLSMs is selected in an empirical manner. Systematic methods for fault threshold definition are based on the statistical properties of the output data which are measured from these electric machines. These methods enable the in-time diagnosis of failures as well as the detection and isolation of incipient faults [18], [19], [269].

In the present chapter, a model-free approach for fault diagnosis of PMLSMs is developed. This method relies on neural modeling of the PMLSM's dynamics and on the use of a statistical decision making procedure for detecting the existence of a parametric change and a failure. To develop the dynamic model of such a type of electric machines, data sets are generated consisting of input-output measurements recorded from its functioning at different operating conditions. The neural model comprises (i) a hidden layer of basis functions having the form of Gauss-Hermite polynomials and (ii) a weights output layer [88], [247], [267], [187]. The neural network is trained with the use of input-output data obtained from the functioning of the PMLSM in the fault-free mode. Weights' adaptation and learning takes place in the form of a first-order gradient algorithm. The approximation error is minimized. The neural model that is obtained through this learning procedure is considered to represent the fault-free functioning of the PMLSM.

The recorded outputs are subtracted from the outputs provided by the neural model. Thus the residuals' sequence is generated. The residuals's data set undergoes statistical processing. It is shown that the sum of the squares of the residuals' vector, being multiplied by the inverse of the associated covariance matrix, stands for a stochastic variable which follows the χ^2 distribution. Next, with the use of the 96% or the 98% confidence intervals of the χ^2 distribution one can define fault thresholds which designate with significant accuracy and undebatable certainty the appearance of a fault in the PMLSM. As long as the value of the aforementioned stochastic variable is within the upper and the lower bound of the confidence interval, it can be concluded that the functioning of the PMLSM is normal. On the other side, whenever the previously noted bounds are exceeded it can be concluded that the functioning of PMLSM has been subjected to fault. Moreover, by processing data from different outputs of the PMLSM, and by repeating the statistical test in subspaces of the state-space model one can also achieve fault isolation which reveals the specific component of the PMLSM that has been affected by the failure (for instance failure in its mechanical part of the machine or failure in its electrical part).

In the following, a model-free neural network-based fault diagnosis method will be analyzed. This approach can be applied to a wide class of dynamical systems. Electric machines such as motors and power generators as well as power electronics such as converters and inverters which are used in electric traction systems and in the smart grid are promising application areas for the proposed fault detection and isolation scheme. It is noted that this specific FDI method is one of the very few solving he problem of fault threshold definition, thus ensuring the detection of incipient failures and the minimization of false alarms. So far, there have been few results on neural networks-based condition monitoring of electric machines [141], [13], [50], [97], [51]. Neural networks can be used to model the dynamics of such systems and subsequently for developing fault diagnosis tools [76], [35], [33], [232], [282]. Usually, the statistics of the monitored signals are not taken into account and statistical criteria for defining fault thresholds are neglected. Unlike this, the present chapter proposes a systematic approach for defining fault thresholds, which is based on the statistical properties of the χ^2 distribution of the signals measured out of the monitored system.

The chapter's developments are meaningful for improving the safety and reliability of electric traction systems and particularly of the traction of Maglev-type electric trains. . Apart from linear motors, the presented method can be also applied to

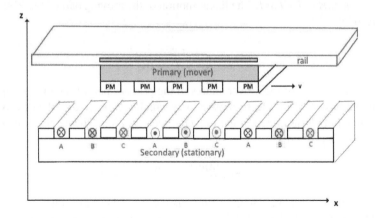

Figure 1.10   Diagram of windings and permanent magnets placement in a PMLSM

rotating electric motors used in electric vehicles. Existing methods for the modeling of electric motors with the use of neural networks do not use activation functions with multi-frequency content, thus not being able to represent efficiently vibrations that these machines may exhibit. Unlike this, the present chapter uses neural networks with Gauss-Hermite activation functions to model PMLSMs. As it holds for wavelet networks, Gauss-Hermite neural networks can also model with high precision the dynamics of electric motors under vibrations or changing operating conditions. Existing methods for fault diagnosis of electric machines often neglect the problem of fault threshold selection and rely on heuristics in decision-taking about the existence of failures. Unlike this, the present chapter proposes a systematic approach for defining fault thresholds, which is based on the statistical properties of the χ^2 distribution of the signals measured out of the monitored system. The proposed method has certainty levels for correct decision taking about the existence of a fault of the order of 96% to 98%.

1.2.2   DYNAMIC MODEL OF THE PMLSM

The Permanent Magnet Linear Synchronous Motor comprises a primary (moving) part where permanent magnets are attached and a secondary (stationary) part where three-phase windings are sequentially arranged (Fig. 1.10). Due to placing the permanent magnets in the magnetic field that is generated by the windings of the secondary part, a propulsion force is exerted on the primary part of the motor [134]. The dynamic model of the PMLSM comprises a mechanical and an electrical part 200.

Mechanical part of PMLSM: The linear motion of the moving part of the PMLSM is described by:

$$\dot{x} = v$$
$$\dot{v} = \frac{1}{M+M_l}[F_e - Dv] \qquad (1.33)$$

where M is the mass of the mover, that is of the primary part of the motor, M_L is the mass of the load, F_e is the electromotive force which is generated by the PMLSM and Dv is a damping term affecting the mover's linear motion. The electromotive force of the motor is given by

$$F_e = 3\pi n_p \frac{[\lambda_{PM} i_q + (L_d - L_q) i_d i_q]}{2\tau} \qquad (1.34)$$

where λ_{PM} is the permanent magnet flux linkage, λ_d is the d-axis flux at the primary part of the motor, λ_q is the q-axis flux at the primary part of the motor, i_d is the d-axis current at the secondary part of the motor and i_q is the q-axis current at the secondary part of the motor. Moreover, L_d is the d-axis inductance at the secondary part of the motor, L_q is the q-axis inductance at the secondary part of the motor, and τ is the poles' pitch.

Electrical part of PMLSM: By applying Kirchhof's law to the secondary part of the motor one has:

$$v_d = i_d R_s + \frac{d}{dt}\lambda_d - \omega_e \lambda_q \qquad (1.35)$$

$$v_q = i_q R_s + \frac{d}{dt}\lambda_q + \omega_e \lambda_d \qquad (1.36)$$

$\omega_e = n_p \frac{\pi v}{\tau}$, n_p is the number of poles, τ is the poles' pitch and v is the linear velocity of the motor. The developed electromagnetic power is

$$P_e = F_e v_e \qquad (1.37)$$

where F_e is the electromotive force defined previously in Eq. (1.34) and $v_e = n_p v$. Thus one has

$$P_e = 3\pi n_p \frac{[\lambda_{PM} i_q + (L_d - L_q) i_d i_q]}{2\tau} n_p v \qquad (1.38)$$

About the magnetic flux of the motor the following relations hold:

$$\lambda_q = L_q i_q$$
$$\lambda_d = L_d i_d + \lambda_{PM} \qquad (1.39)$$

where L_d, L_q are the dq-axes inductances, and λ_{PM} is the permanent magnet flux linkage. Using the above, the state-space description of the PMLSM dynamics is given by

$$\dot{x} = v \qquad (1.40)$$

$$\dot{v} = \frac{1}{(M+M_L)}\left\{3\pi n_p \frac{\lambda_{PM}i_q + (L_d - L_q)i_d i_q}{2\tau} - Dv\right\} \tag{1.41}$$

$$\frac{d}{dt}i_q = -i_q\frac{R_s}{L_q} - \frac{n_p\pi v}{L_q\tau}(L_d i_d + \lambda_{pm}) + \frac{1}{L_q}v_q \tag{1.42}$$

$$\frac{d}{dt}i_d = -i_d\frac{R_s}{L_d} - \frac{n_p\pi v}{L_d\tau}L_q i_q + \frac{1}{L_d}v_d \tag{1.43}$$

Next, by defining the state vector of the PMLSM as $x = [x_1, x_2, x_3, x_4]^T = [x, v, i_q, i_d]^T$ and the control inputs vector as $u = [u_1, u_2]^T = [v_q, v_d]^T$, the state-space description of the system is described by

$$\dot{x}_1 = x_2 \tag{1.44}$$

$$\dot{x}_2 = \frac{1}{(M+M_L)}\left\{3\pi n_p \frac{\lambda_{PM}x_3 + (L_d - L_q)x_4 x_3}{2\tau} - Dv\right\} \tag{1.45}$$

$$\dot{x}_3 = -x_3\frac{R_s}{L_q} - \frac{n_p\pi v}{L_q\tau}(L_d x_4 + \lambda_{pm}) + \frac{1}{L_q}u_1 \tag{1.46}$$

$$\dot{x}_4 = -x_4\frac{R_s}{L_d} - \frac{n_p\pi v}{L_d\tau}L_q x_3 + \frac{1}{L_d}u_2 \tag{1.47}$$

The state-space model of the PMLSM can be also written in the following nonlinear affine-in the-input state-space form:

$$\begin{pmatrix}\dot{x}_1 \\ \dot{x}_2 \\ \dot{x}_3 \\ \dot{x}_4\end{pmatrix} = \begin{pmatrix} x_2 \\ \frac{1}{(M+M_L)}\left\{3\pi n_p\frac{\lambda_{pm}x_3 + (L_d - L_q)x_3 x_4}{2\tau}\right\} - Dx_2 \\ -\frac{R_s}{L_q}x_3 - \frac{n_p\pi}{L_q\tau}L_d x_2 x_4 - \frac{n_p\pi}{L_q\tau}\lambda_{pm}x_2 \\ -\frac{R_s}{L_q}x_4 - \frac{n_p\pi}{L_q\tau}L_d x_2 x_3\end{pmatrix} + \begin{pmatrix} 0 & 0 \\ 0 & 0 \\ \frac{1}{L_q} & 0 \\ 0 & \frac{1}{L_d}\end{pmatrix}\begin{pmatrix}u_1 \\ u_2\end{pmatrix}$$

$$\tag{1.48}$$

The system's state-space description can be also written in the concise form

$$\dot{x} = f(x) + G(x)u \tag{1.49}$$

where $x \in R^{4\times 1}$, $u \in R^{2\times 1}$, $f(x) \in R^{4\times 1}$ and $G(x) \in R^{4\times 2}$.

1.2.3   MODELING OF THE PMLSM WITH THE USE OF NEURAL NETWORKS

1.2.3.1   Feed-forward neural networks for nonlinear systems modeling

The proposed fault diagnosis approach for PMLSMs that exhibit nonlinear dynamics, can be implemented with the use of feed-forward neural networks. The idea of function approximation with the use of feed-forward neural networks (FNN) comes from generalized Fourier series. It is known that any function $\psi(x)$ in a L^2 space can be expanded, using generalized Fourier series in a given orthonormal basis, i.e. [190]

$$\psi(x) = \sum_{k=1}^{\infty} c_k \psi_k(x), \ a \leq x \leq b \tag{1.50}$$

Truncation of the series yields in the sum

$$S_M(x) = \sum_{k=1}^{M} a_k \psi_k(x) \tag{1.51}$$

If the coefficients a_k are equal to the generalized Fourier coefficients, i.e. when $a_k = c_k = \int_a^b \psi(x)\psi_k(x)dx$, then Eq. (1.51) is a mean square optimal approximation of $\psi(x)$.

Unlike generalized Fourier series, in FNN the basis functions are not necessarily orthogonal. The hidden units in a FNN usually have the same activation functions and are often selected as sigmoidal functions or Gaussians. A typical feed-forward neural network consists of n inputs x_i, $i = 1, 2, \cdots, n$, a hidden layer of m neurons with activation function $h : R \to R$ and a single output unit (see Fig. 1.11(a)). The FNN's output is given by

$$\psi(x) = \sum_{j=}^{n} c_j h(\sum^{n} w_{ji} x_i + b_j) \tag{1.52}$$

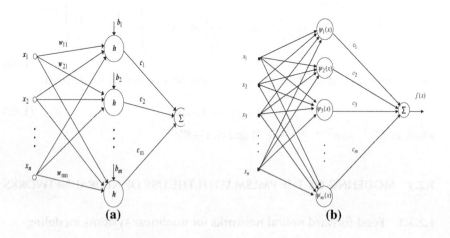

(a)   **(b)**

Figure 1.11   (a) Feed-forward neural network and (b) Neural network with Gauss-Hermite basis functions

The root mean square error in the approximation of function $\psi(x)$ by the FNN is given by

$$E_{RMS} = \sqrt{\frac{1}{N}\sum_{k=1}^{N}(\psi(x^k) - \hat{\psi}(x^k))^2} \tag{1.53}$$

where $x^k = [x_1^k, x_2^k, \cdots, x_n^k]$ is the k-th input vector of the neural network. The activation function is usually a sigmoidal function $h(x) = \frac{1}{1+e^{-x}}$ while in the case of radial basis functions networks it is a Gaussian [187]. Several learning algorithms for neural networks have been studied. The objective of all these algorithms is to find numerical values for the network's weights so as to minimize the mean square error E_{RMS} of Eq. (1.53). The algorithms are usually based on first and second order gradient techniques. These algorithms belong to: i) batch-mode learning, where to perform parameters update the outputs of a large training set are accumulated and the mean square error is calculated (back-propagation algorithm, Gauss-Newton method, Levenberg-Marquardt method, etc.), ii) pattern-mode learning, in which training examples are run in cycles and the parameters update is carried out each time a new datum appears (Extended Kalman Filter algorithm) [189].

Unlike conventional FNN with sigmoidal or Gaussian basis functions, Hermite polynomial-based FNN remain closer to Fourier series expansions by employing activation functions which satisfy the property of orthogonality [187]. Other basis functions with the property of orthogonality are Hermite, Legendre, Chebyshev, and Volterra polynomials [187].

1.2.3.2 Neural Networks using Gauss-Hermite activation functions

The Gauss-Hermite series expansion

Next, as orthogonal basis functions of the feed-forward neural network Gauss-Hermite activation functions are considered [198], [190]:

$$X_k(x) = H_k(x)e^{\frac{-x^2}{2}}, \; k = 0, 1, 2, \cdots \tag{1.54}$$

where $H_k(x)$ are the Hermite orthogonal functions (Fig. 1.12). The Hermite functions $H_k(x)$ are also known to be the eigenstates of the quantum harmonic oscillator. The general relation for the Hermite polynomials is

$$H_k(x) = (-1)^k e^{x^2} \frac{d^{(k)}}{dx^{(k)}} e^{-x^2} \tag{1.55}$$

According to Eq. (1.55) the first five Hermite polynomials are:

$H_0(x) = 1, H_1(x) = 2x, H_2(x) = 4x^2 - 2, H_3(x) = 8x^3 - 12x, H_4(x) = 16x^4 - 48x^2 + 12$

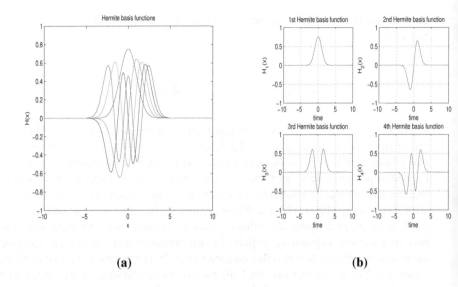

(a) **(b)**

Figure 1.12 (a) First five one-dimensional Hermite basis functions and (b) analytical representation of the 1D Hermite basis function

It is known that Hermite polynomials are orthogonal, i.e. it holds

$$\int_{-\infty}^{+\infty} e^{-x^2} H_m(x) H_k(x) dx = \begin{cases} 2^k k! \sqrt{\pi} \; if \; m = k \\ 0 \; if \; m \neq k \end{cases} \tag{1.56}$$

Using now, Eq. (1.56), the following basis functions can be defined 187:

$$\psi_k(x) = [2^k \pi^{\frac{1}{2}} k!]^{-\frac{1}{2}} H_k(x) e^{-\frac{x^2}{2}} \tag{1.57}$$

where $H_k(x)$ is the associated Hermite polynomial. From Eq. (1.56), the orthogonality of basis functions of Eq. (1.57) can be concluded, which means

$$\int_{-\infty}^{+\infty} \psi_m(x) \psi_k(x) dx = \begin{cases} 1 \; if \; m = k \\ 0 \; if \; m \neq k \end{cases} \tag{1.58}$$

Moreover, to achieve multi-resolution analysis Gauss-Hermite basis functions of Eq. (1.57) are multiplied with the scale coefficient α. Thus the following basis functions are derived [187]

$$\beta_k(x, \alpha) = \alpha^{-\frac{1}{2}} \psi_k(\alpha^{-1} x) \tag{1.59}$$

which also satisfy orthogonality condition

$$\int_{-\infty}^{+\infty} \beta_m(x, \alpha) \beta_k(x, \alpha) dx = \begin{cases} 1 \; if \; m = k \\ 0 \; if \; m \neq k \end{cases} \tag{1.60}$$

Any function $f(x)$, $x \in R$ can be written as a weighted sum of the above orthogonal basis functions, i.e.

$$f(x) = \sum_{k=0}^{\infty} c_k \beta_k(x, \alpha) \tag{1.61}$$

where coefficients c_k are calculated using the orthogonality condition

$$c_k = \int_{-\infty}^{+\infty} f(x)\beta_k(x, \alpha)dx \tag{1.62}$$

Assuming now that instead of infinite terms in the expansion of Eq. (1.61), M terms are maintained, then an approximation of $f(x)$ is achieved. The expansion of $f(x)$ using Eq. (1.61) is a Gauss-Hermite series. Eq. (1.61) is a form of Fourier expansion for $f(x)$. Eq. (1.61) can be considered as the Fourier transform of $f(x)$ subject only to a scale change. Indeed, the Fourier transform of $f(x)$ is given by

$$F(s) = \frac{1}{2\pi} \int_{-\infty}^{+\infty} f(x)e^{-jsx}dx \Rightarrow f(x) = \frac{1}{2\pi} \int_{-\infty}^{+\infty} F(s)e^{jsx}ds \tag{1.63}$$

The Fourier transform of the basis function $\psi_k(x)$ of Eq. (1.57) satisfies [187]

$$\Psi_k(s) = j^k \psi_k(s) \tag{1.64}$$

while for the basis functions $\beta_k(x, \alpha)$ using scale coefficient α, it holds that

$$B_k(s, \alpha) = j^k \beta_k(s, \alpha^{-1}) \tag{1.65}$$

Therefore, it holds

$$f(x) = \sum_{k=0}^{\infty} c_k \beta_k(x, \alpha) \xrightarrow{F} F(s) = \sum_{k=0}^{\infty} c_k j^n \beta_k(s, \alpha^{-1}) \tag{1.66}$$

which means that the Fourier transform of Eq. (1.61) is the same as the initial function , subject only to a change of scale. The structure of a a feed-forward neural network with Hermite basis functions is depicted in Fig. 1.11(b).

Neural Networks using 2D Hermite activation functions

Two-dimensional Hermite polynomial-based neural networks can be constructed by taking products of the one dimensional basis functions $B_k(x, \alpha)$. Thus, setting $x = [x_1, x_2]^T$ one can define the following basis functions [198], [190]

$$B_k(x, \alpha) = \frac{1}{\alpha} B_{k_1}(x_1, \alpha)B_{k_2}(x_2, \alpha) \tag{1.67}$$

These two dimensional basis functions are again orthonormal, i.e. it holds

$$\int d^2x B_n(x, \alpha)B_m(x, \alpha) = \delta_{n_1 m_1}\delta_{n_2 m_2} \tag{1.68}$$

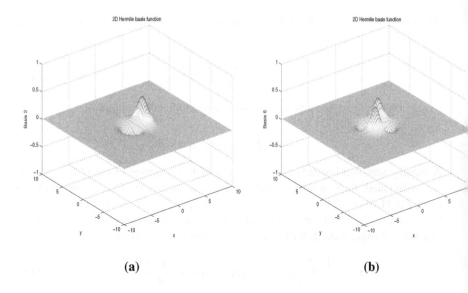

(a)   **(b)**

Figure 1.13   2D Hermite polynomial activation functions: (a) basis function $B_2(x, \alpha)$ and (b) basis function $B_6(x, \alpha)$

The basis functions $B_k(x)$ are the eigenstates of the two-dimensional harmonic oscillator and form a complete basis for integrable functions of two variables. A two dimensional function $f(x)$ can thus be written in the series expansion:

$$f(x) = \sum_{k_1, k_2}^{\infty} c_k B_k(x, \alpha) \tag{1.69}$$

The choice of an appropriate scale coefficient α and of maximum order k_{max} is of practical interest. The coefficients c_k are given by

$$c_k = \int dx^2 f(x) B_k(x, \alpha) \tag{1.70}$$

Indicative basis functions $B_2(x, \alpha)$, $B_6(x, \alpha)$, $B_9(x, \alpha)$, $B_{11}(x, \alpha)$ and $B_{13}(x, \alpha)$, $B_{15}(x, \alpha)$ of a 2D feed-forward neural network with Hermite basis functions are depicted in Fig. 1.13, Fig. 1.14, and Fig. 1.15. Following, the same method N-dimensional Hermite polynomial-based neural networks ($N > 2$) can be constructed. The associated high-dimensional Gauss-Hermite activation functions preserve the properties of orthogonality and invariance to Fourier transform.

1.2.4   STATISTICAL FAULT DIAGNOSIS USING THE NEURAL NETWORK

1.2.4.1   Fault detection

A Gauss-Hermite neural network has been used to learn the dynamics of the PMLSM. For each functioning mode, the training set comprised $N = 4000$ vectors

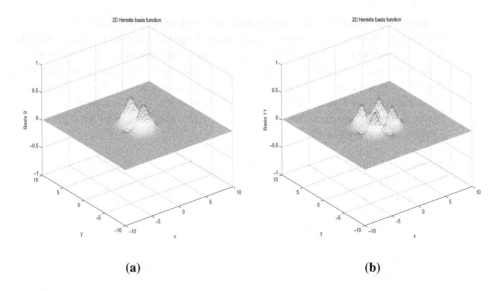

(a)   **(b)**

Figure 1.14   2D Hermite polynomial activation functions: (a) basis function $B_9(x, \alpha)$ and (b) basis function $B_{11}(x, \alpha)$

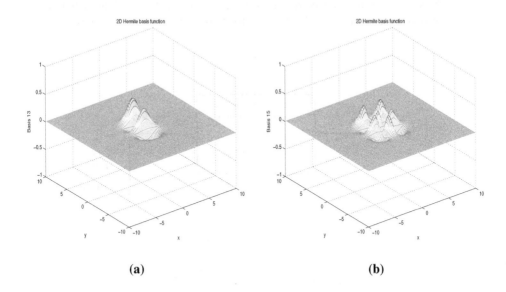

(a)   **(b)**

Figure 1.15   2D Hermite polynomial activation functions: (a) basis function $B_{13}(x, \alpha)$ and (b) basis function $B_{15}(x, \alpha)$

of input-output data. In each vector the input data were $x_1(k)$, $x_1(k-1)$, $x_1(k-2)$ standing for the three most recent values of the first state variable x_1 of the PMLSM. It was considered that each input variable can be expressed in a series expansion form using the first four Gauss-Hermite basis functions. The output of the neural network was the estimated value of first state variable $\hat{x}_1(k+1)$. Following the previous concept, the Gauss-Hermite neural network which has been used for learning the input-output data comprised $4^3 = 64$ basis functions in its hidden layer and 64 weights in its output layer.

The residuals' sequence, that is the differences between (i) the real outputs of the PMLSM and (ii) the outputs estimated by the neural network (Fig. 1.16) is a discrete error process e_k with dimension $m \times 1$, here $m = N$ is the dimension of the output measurements vector. Actually, it is a zero-mean Gaussian white-noise process with covariance given by E_k 198.

Following again the stages which were described in the previous section, a conclusion can be stated based on a measure of certainty that the PMLSM has not been subjected to a fault. To this end, the following *normalized error square* (NES) is defined 198

$$\varepsilon_k = e_k^T E_k^{-1} e_k \tag{1.71}$$

The sum of this normalized residuals' square follows a χ^2 distribution, with a number of degrees of freedom that is equal to the dimension of the residuals' vector. An

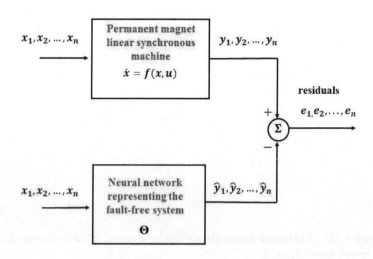

Figure 1.16   Residuals' generation for the PMLSM, with the use of a neural network

appropriate test for the normalized error sum is to numerically show that the following condition is met within a level of confidence (according to the properties of the χ^2 distribution)

$$E\{\varepsilon_k\} = m \qquad (1.72)$$

This can be achieved using statistical hypothesis testing, which is associated with confidence intervals. A 95% confidence interval is frequently applied, which is specified using the probability region $100(1 - a)$ with $a = 0.05$. Actually, a two-sided probability region is considered cutting-off two end tails of 2.5% each. For M runs the normalized error square that is obtained is given by

$$\bar{\varepsilon}_k = \frac{1}{M} \sum_{i=1}^{M} \varepsilon_k(i) = \frac{1}{M} \sum_{i=1}^{M} e_k^T(i) E_k^{-1}(i) e_k(i) \qquad (1.73)$$

where ε_i stands for the i-th run at time t_k. Then $M\bar{\varepsilon}_k$ will follow a χ^2 density with Mm degrees of freedom. This condition can be checked using a χ^2 test. The hypothesis holds, if the following condition is satisfied

$$\bar{\varepsilon}_k \in [\zeta_1, \zeta_2] \qquad (1.74)$$

where ζ_1 and ζ_2 are derived from the tail probabilities of the χ^2 density.

1.2.4.2 Fault isolation

By applying the statistical test into the individual components of the PMLSM, it is also possible to find out the specific component that has been subjected to a fault [198]. For a PMLSM model of n parameters suspected for change one has to carry out n χ^2 statistical change detection tests, where each test is applied to the subset that comprises parameters $i - 1$, i and $i + 1$, $i = 1, 2, \cdots, n$. Actually, out of the n χ^2 statistical change detection tests, the ones that exhibit the highest score are those that identify the parameter that has been subjected to change. In the case of multiple faults one can identify the subset of parameters that has been subjected to change by applying the χ^2 statistical change detection test according to a combinatorial sequence. This means that

$$\binom{n}{k} = \frac{n!}{k!(n-k)!} \qquad (1.75)$$

tests have to take place, for all clusters in the PMLSM's model, that finally comprise n, $n - 1$, $n - 2$, $\cdots$, 2, 1 parameters. Again the χ^2 tests that give the highest scores indicate the parameters which are most likely to have been subjected to change.

As a whole, the concept of the proposed fault detection and isolation method is simple. The sum of the squares of the residuals' vector, weighted by the inverse of the residuals' covariance matrix, stands for a stochastic variable which follows the χ^2

distribution. Actually, this is a multi-dimensional χ^2 distribution and the number of its degrees of freedom is equal to the dimension of the residuals' vector. Since, there is 1 measurable output of the PMLSM, the residuals' vector is of dimension 1 and the number of degrees of freedom is also 1. Next, from the properties of the χ^2 distribution, the mean value of the aforementioned stochastic variable in the fault-free case should be also 1. However, due to having sensor measurements affected by noise, the value of the statistical test in the fault-free case will not be precisely equal to 1 but it may vary within a small range around this value. This range is determined by the confidence intervals of the χ^2 distribution. For a probability value of 98% to get a value of the stochastic variable about 1, the associated confidence interval is given by the lower bound $L = 0.936$ and by the upper bound $U = 1.066$. Consequently, as long as the statistical test provides an indication that the aforementioned stochastic variable is in the interval $[L, U]$ the functioning of the PMLSM can be concluded to be free of faults. On the other side, when the bounds of the previously given confidence interval are exceeded it can be concluded that the PMLSM has undergone a fault. Finally, by performing the statistical test in subspaces of the PMLSM's state-space model, where each subspace is associated with different components, one can also achieve fault isolation. This signifies that the specific component that has caused the malfunctioning of the PMLSM can be identified.

1.2.5   SIMULATION TESTS

The performance of the proposed fault diagnosis method for the PMLSM has been tested through simulation experiments. Six different functioning modes have been considered. These are depicted in Fig 1.17 to Fig. 1.19.

The training of the Gauss-Hermite neural network which has been used to model the dynamics of the PMLSM was performed in Matlab. The fault diagnosis algorithm for the PMLSM which has been based on the confidence intervals of the χ^2 distribution has been also developed in Matlab. The induced changes and disturbances in the PMLSM where incipient and as confirmed by the previously noted diagrams they can be hardly distinguished by human supervisors of the system. By applying the previously analyzed statistical test which relies on the properties of the χ^2 distribution, results have been obtained about the detection of failures in the PMLSM. These results are depicted in Fig. 1.20 to Fig. 1.25. Actually, two types of fault detection tests are presented: (i) consecutive statistical tests carried out on the PMLSM and (ii) mean value of the statistical tests.

The number of monitored outputs in the PMLSM was $n = 1$. Consequently, according to the previous analysis, in the fault-free case the mean value of the statistical test should be very close to the value 1. In particular considering 96% confidence intervals, the bounds of the normal functioning of the system were $U = 1.056$ (upper

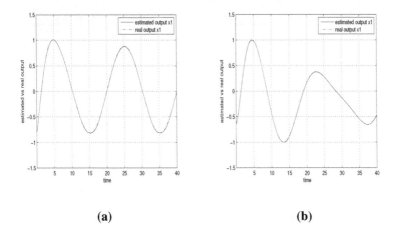

(a) **(b)**

Figure 1.17 (a) test case 1 (tracking of setpoint 1 and fault at the output of the PMLSM): Real value (blue) of the PMLSM's output x_1 vs estimated value $\hat{x}_1$ (*red*) and (b) test case 2 (tracking of setpoint 2 and fault at the resistance R_s of the windings of the PMLSM)): Real value (blue) of the PMLSM's output x_1 vs estimated value $\hat{x}_1$ (red)

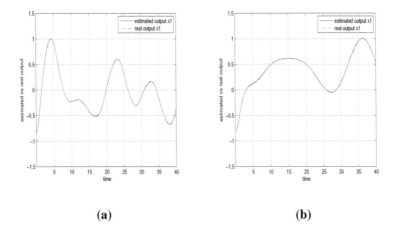

(a) **(b)**

Figure 1.18 (a) test case 3 (tracking of setpoint 3 and fault at the output of the PMLSM): Real value (blue) of the PMLSM's output x_1 vs estimated value $\hat{x}_1$ (red) and (b) test case 4 (tracking of setpoint 4 and fault at the resistance R_s of the windings of the PMLSM): Real value (blue) of the PMLSM's output x_1 vs estimated value $\hat{x}_1$ (red)

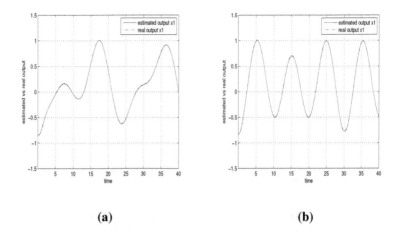

(a)   **(b)**

Figure 1.19   (a) test case 5 (tracking of setpoint 5 and fault at the output of the PMLSM): Real value (blue) of the PMLSM's output x_1 vs estimated value $\hat{x}_1$ (red) and (b) test case 6 (tracking of setpoint 6 and fault at the resistance R_s of the windings of the PMLSM): Real value (blue) of the PMLSM's output x_1 vs estimated value $\hat{x}_1$

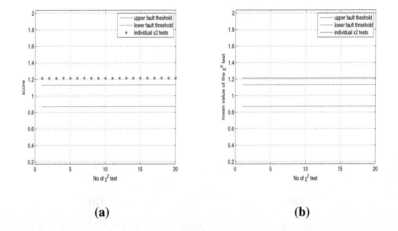

(a)   **(b)**

Figure 1.20   test case 1: fault (additive disturbance) at the output of the PMLSM (a) individual χ^2 tests and related confidence intervals and (b) mean value of the χ^2 test and related confidence intervals

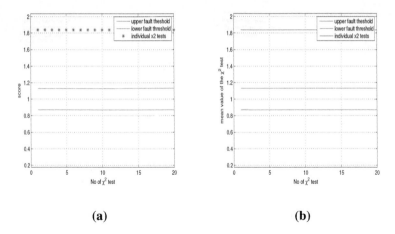

(a) **(b)**

Figure 1.21 test case 2: fault (parametric change) at the resistance R_s of the windings of the PMLSM (a) individual χ^2 tests and related confidence intervals and (b) mean value of the χ^2 test and related confidence intervals

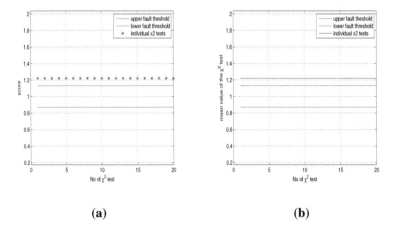

(a) **(b)**

Figure 1.22 test case 3: fault (additive disturbance) at the output of the PMLSM (a) individual χ^2 tests and related confidence intervals and (b) mean value of the χ^2 test and related confidence intervals

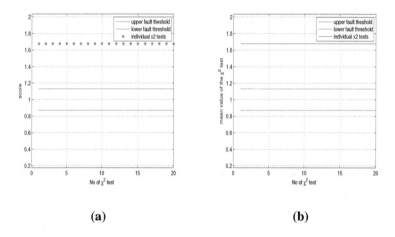

Figure 1.23 Test case 4: fault (parametric change) at the resistance R_s of the windings of the PMLSM (a) individual χ^2 tests and related confidence intervals and (b) mean value of the χ^2 test and related confidence intervals

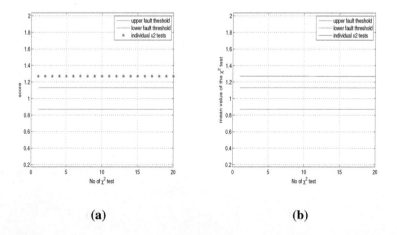

Figure 1.24 Test case 5: fault (additive disturbance) at the output of the PMLSM (a) individual χ^2 tests and related confidence intervals and (b) mean value of the χ^2 test and related confidence intervals

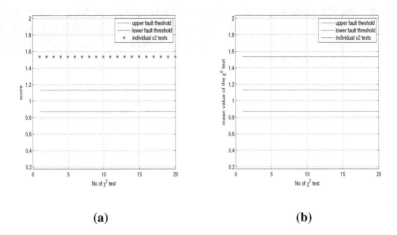

(a)　　　　　　　　　　　　　　　　(b)

Figure 1.25　Test case 6: fault (parametric change) at the resistance R_s of the windings of the PMLSM (a) individual χ^2 tests and related confidence intervals and (b) mean value of the χ^2 test and related confidence intervals

bound) and $L = 0.945$ (lower bound). As long as the mean value of the statistical test remains within the bounds of this confidence interval, it can be concluded that the functioning of the PMLSM is normal. On the other side, whenever the value of the statistical test exceeded persistently the aforementioned bounds, then one could infer the appearance of a parametric change or of a strong external perturbation. This is shown in all cases 1 to 6. Finally, it is noted that by performing the statistical test in subspaces of its state-space model one can also achieve fault isolation. For instance, it is possible to conclude the existence of a failure in the mechanical part of the PMLSM, or a failure in the electrical part of this electric machine.

An issue that arises is why one to use the χ^2 test for fault diagnosis instead of the popular p-value test. As far as the p-value test is concerned this can be performed on the stochastic value which is obtained from the residuals' data, after using the Central Limit Theorem (CLT). The p-value test is related with the probability, for obtaining a value of the test statistic $z = (\bar{x} - \mu)/(\sigma/\sqrt{n})$ that annihilates the null hypothesis (where $\bar{x}$ is the average value, μ is the samples' mean, σ is the sample standard deviation, and n is the sample's size. This is equivalent to saying that one uses the $a\%$ confidence intervals of the χ^2 distribution (for instance with a to be 96% or 98%), because in the latter case the probability for an erroneous conclusion from the χ^2 test is $1 - a\%$ (for instance 4% or 2% respectively). Consequently, there is no benefit from using a p-value test for finding faults in the monitored system. On the other side, the p-value test is not efficient in processing a multi-dimensional residuals vector, whereas the χ^2 test is. One has to apply the p-value test for the sequence of the individual elements of the residuals vector of the monitored system, whereas in the χ^2 test the sequence of the entire residuals vector can be processed in batch mode. Additionally, it is noted that the application of the p-value test, requires to know the mean value and the variance of the i.i.d. processed data distribution (μ and σ) whereas the χ^2 test does not.

2 Fault diagnosis for SG-based renewable energy systems

2.1 CONTROL OF THE MARINE-TURBINE AND SYNCHRONOUS-GENERATOR UNIT

2.1.1 OUTLINE

In this chapter the control for a marine-turbine and synchronous-generator power unit is treated first. The need for transition to carbon-free electric power generation has intensified research on renewable energy systems [198], [44]. Ocean energy resources are estimated to be sufficient for covering the entire world's energy consumption [44]. They account for 2 million TWh per year, while out of them up to 92000TWh are technically exploitable. In particular the technically exploitable wave energy raises to 32000TWh per year, while the technically exploitable tidal energy raises to 26000TWh per year [44]. At present, the level of real exploitation of marine power remains low and is amenable to significant improvement. The economically viable marine power installations provide annually only 180TWh [44]. Comparing to wind power, marine power is predictable and its production levels exhibit less fluctuations. Sea currents at a speed which is at 10% of wind speed, enable marine power units reach the same power production as wind power units. This allows for covering more efficiently power consumption needs. On annual basis and during normal operating conditions, a marine power unit can reach 40-50% of its maximum capacity for energy supply, while the equivalent ratings for wind power units do not exceed 25-30% [44].

The grid integration of marine power units requires the development of power transmission and distribution infrastructure, including power electronics such as converters and inverters being controlled with elaborated control techniques [170], [106], [285], [173]. This allows for synchronizing marine power units with the frequency of the main electricity grid, operating these power units at variable speed for more efficient energy harvesting, as well as reaching the targeted power ratings despite variations in the excitation that is provided by sea currents [56], [57], [70]. Besides, control systems knowhow allows for preserving the reliable functioning of marine power units despite failure of equipment, grid faults or other external perturbations [46], [174], [289]. Horizontal and vertical axis turbines are the most frequently met in marine power generation [246], [135], [82]. Synchronous generators of various types are used frequently in marine power units (such as synchronous generators with windings at both the stator and the rotor, permanent magnet synchronous gen-

DOI: 10.1201/9781003527657-2

erators, reluctance generators, and multi-phase synchronous generators) [286], [24], [45]. Besides, asynchronous generators are also used in marine power units (such as doubly-fed induction generators) [25], [77].

Nonlinear control for marine power generation units remains a nontrivial and challenging task [238], [290], [287]. Efficient nonlinear control methods have to compensate for the nonlinear and multi-variable dynamics of the integrated marine turbine and electric power generator system [55], [234], [92]. In the present chapter, a new nonlinear optimal control method is developed for the dynamic model of the marine-turbine and synchronous power generator [198]. The method performs first an approximate linearization of the state-space model of the marine power unit with the use of first-order Taylor series expansion and through the computation of the associated Jacobian matrices [188], [18], [189]. The linearization point is updated at each sampling instance and is defined by the present value of the system's state vector and by the last sampled value of the control inputs vector. For the approximately linearized model of the marine power generation unit a stabilizing H-infinity controller is designed. To compute the controller's feedback gains an algebraic Riccati equation is repetitively solved at each time step of the control algorithm [202], [195], [191], [204]. The global stability and the robustness properties of the control method are proven through Lyapunov analysis.

Next, the novel nonlinear optimal control method is compared against other nonlinear control techniques one could have considered for marine power units that comprise a horizontal axis turbine and a synchronous generator. The nonlinear optimal control method is compared against [198]: (i) Lie algebra-based control, (ii) flatness-based control with transformation into canonical forms, (iii) flatness-based control implemented in successive loops, (iv) sliding-mode control, (v) multiple models-based fuzzy control and (vi) PID control. The chapter's theoretical analysis and experimental results demonstrate that: (a) the presented nonlinear optimal control method has improved performance when compared against other nonlinear control schemes that one can consider for the dynamic model of marine-turbine power generators, (b) it achieves fast and accurate tracking of all reference setpoints for the marine power unit under moderate variations of the control inputs and (c) it minimizes the dispersion of energy by the control system of the marine power unit.

In particular, the comparison of the proposed nonlinear optimal (H-infinity) control method against the aforementioned nonlinear control schemes for the marine-turbine and synchronous generator power unit, shows the following: (1) unlike global linearization-based control approaches, such as Lie algebra-based control and differential flatness theory-based control, the optimal control approach does not rely on complicated transformations (diffeomorphisms) of the system's state variables. Besides, the computed control inputs are applied directly on the initial nonlinear model of the marine-power unit and not on its linearized equivalent description. The inverse transformations which are met in global linearization-based control are avoided and consequently one does not come against the related singularity problems. (2) unlike sliding-mode control and backstepping control, the proposed optimal control method does not require the state-space description of the system to be found in a specific form. About sliding-mode control, it is known that when the controlled system is not found in the input-output linearized form the definition of the sliding surface can be an intuitive procedure. About backstepping-type control, it is known that it can not be directly applied to a dynamical system if the related state-space model is not found in the triangular (backstepping integral) form, (3) unlike multiple local models-based control, the nonlinear optimal control method uses only one linearization point and needs the solution of only one Riccati equation so as to compute the stabilizing feedback gains of the controller. Consequently, in terms of computation load the proposed control method for marine-turbine power units is much more efficient (4) unlike PID control, the proposed nonlinear optimal control method is of proven global stability, the selection of the controller's parameters does not rely on a heuristic tuning procedure, and the stability of the control loop is assured in the case of changes of operating points (5) unlike approaches for optimal control met in industry, such as Model Predictive Control (MPC) and Nonlinear Model Predictive control (NMPC), the proposed control method is of proven global stability. It is known that MPC is a linear control approach that if applied to the nonlinear dynamics of the marine turbine power units, the stability of the control loop will be lost. Besides, in NMPC the convergence of its iterative search for an optimum depends on initialization and parameter values selection and consequently the global stability of this control method cannot be always assured.

Table I	
Parameters of the dynamic model of the marine power generation unit	
Parameter	*Definition*
θ_t	Turn angle of the turbine
ω_t	Angular speed of the turbine
θ_g	Turn angle of the synchronous generator
ω_g	Angular speed of the synchronous generator
ω_0	Synchronous (reference) speed of the generator
E_q'	Quadrature axis transient voltage of the generator
T_m	Mechanical torque applied on the turbine
E_f	Excitation voltage of the generator
J_t	Moment of inertia of the marine turbine
K_1	Elasticity of the turbine-generator drivetrain
D_t	Damping coefficient of the turbine's rotation
c_{b_a}	Nonlinear function of the turbine blades' pitch angle
J_g	Moment of inertia of the generator's rotor
D_g	Damping coefficient of the generator's rotation
E_q	Quadrature axis voltage of the generator
V_s	infinite bus (transmission line's) voltage

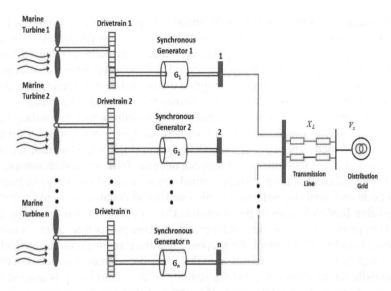

Figure 2.1   Diagram of distributed marine turbine power generation units. Each unit comprises a marine turbine, a drivetrain and a synchronous generator

2.1.2 DYNAMIC MODEL OF THE MARINE TURBINE POWER GENERATION SYSTEM

The marine power unit consists of an horizontal axis marine turbine and of a synchronous generator (Fig. 2.1). Considering that the drivetrain which connects the turbine to the generator comprises: (i) a gear of n_t teeth at the side of the marine-turbine and (ii) a gear of n_g teeth at the side of the synchronous generator, one has that the dynamic model of the power generation unit is given by the following set of differential equations [198]:

$$\begin{pmatrix} \dot{x}_1 \\ \dot{x}_2 \\ \dot{x}_3 \\ \dot{x}_4 \\ \dot{x}_5 \end{pmatrix} = \begin{pmatrix} x_2 \\ \frac{1}{J_t}[-K_1(x_1 - x_3) - Dx_2] \\ x_4 \\ \frac{1}{J_g}[K_1(x_1 - x_3)\frac{n_g}{n_t} - D_gx_4 - \omega_0\frac{V_sx_5sin(x_3)}{x_{d\Sigma}'}] \\ -\frac{1}{T_d'}x_5 + \frac{1}{T_{do}}\frac{x_d - x_d'}{x_{d\Sigma}'}V_scos(x_3) \end{pmatrix} + \begin{pmatrix} 0 & 0 \\ \frac{T_m}{J_t} & 0 \\ 0 & 0 \\ 0 & 0 \\ 0 & \frac{1}{T_{d0}} \end{pmatrix}\begin{pmatrix} u_1 \\ u_2 \end{pmatrix} \quad (2.1)$$

where $x_1 = \theta_t$, $\theta_2 = \omega_t$, $x_3 = \theta_g$, $x_4 = \omega_g$ and $x_5 = E_q'$, while $u_1 = c_{ba}$ and $u_2 = E_f$. The state variables of the model of the marine power generation unit are defined as follows: θ_t is the turn angle of the turbine, θ_g is the turn angle of the generator's rotor, ω_t is the angular speed of the turbine, ω_g is the angular speed of the generator with reference to the synchronous speed ω_0, E_q' is the quadrature axis transient voltage of the generator which actually indicates the magnetic flux of the generator, T_m is the mechanical torque provided by the marine currents and E_f is the excitation voltage.

The main parameters of the marine-turbine and synchronous power generator are outlined in Table I. The dynamics of the of the marine turbine is given by the first two rows of the state-space model:

$$\dot{\theta}_t = \omega_t$$
$$\dot{\omega}_t = \frac{1}{J_t}[-K_1(\theta_1 - \theta_g) - D_t\omega_t] + \frac{T_m}{J_t}c_{ba} \qquad (2.2)$$

In the above equations J_t is the moment of inertia of the marine turbine, K_1 is the elasticity of the drivetrain connecting the turbine with the rotor of the generator which results in torque (torsion) equal to $K_1(\theta_t - \theta_g)$, and D_t is a damping (friction) coefficient for the turn motion of the turbine. Coefficient c_{ba} stands for a control input to the generator and is related primarily with the variable pitch angle of the generator's blades. It regulates the effect of the mechanical (tidal) torque that is exerted on the marine-turbine. Actually c_{ba} is a nonlinear function of the blades' pitch angle, of the rotor's speed and of the tidal stream's speed. This function may change by the water's speed. In practice the value of c_{ba} is part of the hydrodynamic maps of the turbine and is stored in look-up tables. The mechanical torque T_m which is provided to the marine turbine by sea-currents or tidal streams is given by

$$T_m = \frac{1}{2}\rho_i S V_t^3 \qquad (2.3)$$

where ρ is the density of the water, S is the cross-sectional area of the marine turbine and V_t is the velocity of the tidal stream.

The dynamics of the electrical part of the marine power unit, which consists of the synchronous generator and of the power transmission line (Single Machine Infinite Bus model), is analyzed in the following:

$$\dot{\theta}_g = \omega_g$$
$$\dot{\omega}_g = [\frac{1}{J_g}[K_1(\theta_t - \theta_g)]\frac{n_g}{n_t} - D_g\omega_g - \frac{\omega_0 V_s E_q' sin(\theta_g)}{x_{d\Sigma}'}]$$
$$\dot{E}_q' = -\frac{1}{T_d'}E_q' + \frac{1}{T_{do}'}\frac{x_d - x_d'}{x_{d\Sigma}'}V_s cos(\theta_g) + \frac{1}{T_{do}'}E_f \qquad (2.4)$$

where θ_g is the turn angle of the generator's rotor, ω_g is the rotation speed of the rotor with respect to synchronous reference, ω_0 is the synchronous speed of the generator, J_g is the moment of inertia of the rotor, $T_e = -\frac{\omega_0 V_s E_q' sin(\theta_g)}{x_{d\Sigma}'}$ is the electromagnetic torque of the generator which is associated with the produced active power, $T_g = \frac{1}{J_g}[K_1(\theta_t - \theta_g)]\frac{n_g}{n_t}$ is the mechanical input torque to the generator which is provided by the turbine's rotation, and D_g is the damping constant of the generator. Moreover, the following variables are defined: $\Delta\theta_g = \theta_g - \theta_{g0}$ and $\Delta\omega_g = \omega_g - \omega_0$ with ω_0 denoting the synchronous speed. The generator's electrical dynamics is described as follows [198]:

$$\dot{E}_q' = \frac{1}{T_{do}'}(E_f - E_q) \qquad (2.5)$$

where E_q' is the quadrature-axis transient voltage of the generator, E_q is the quadrature axis voltage of the generator, T_{d_o} is the direct axis open-circuit transient time constant of the generator and E_f is the equivalent voltage in the excitation coil. The algebraic equations of the synchronous power generator, in the Single Machine Infinite Bus model, are given by:

$$E_q = \frac{x_{d\Sigma}}{x_{d\Sigma}'} E_q' - (x_d - x_d') \frac{V_s}{x_{d\Sigma}} \cos(\Delta\delta)$$

$$I_q = \frac{V_s}{x_{d\Sigma}'} \sin(\Delta\delta)$$

$$I_d = \frac{E_q'}{x_{d\Sigma}'} - \frac{V_s}{x_{d\Sigma}'} \cos(\Delta\delta)$$

$$P_e = \frac{V_s E_q'}{x_{d\Sigma}'} \sin(\Delta\delta) \tag{2.6}$$

$$Q_e = \frac{V_s E_q'}{x_{d\Sigma}'} \cos(\Delta\delta) - \frac{V_s^2}{x_{d\Sigma}'}$$

$$V_t = \sqrt{(E_q' - X_d' I_d)^2 + (X_d' I_q)^2}$$

where $x_{d\Sigma} = x_d + x_T + x_L$, $x_{d\Sigma}' = x_d' + x_T + x_L$, x_d is the direct-axis synchronous reactance, x_T is the reactance of the transformer, x_d' is the direct axis transient reactance, x_L is the reactance of the transmission line, I_d and I_q are direct and quadrature axis currents of the generator, V_s is the infinite bus voltage, P_e is the generator's active power, Q_e is the generator reactive power delivered to the infinite bus, and V_t is the terminal voltage of the generator. Besides, $T_d' = \frac{x_{d\Sigma}'}{x_{d\Sigma}} T_{d_o}$ is the time constant of the field winding, and E_f is the excitation voltage.

2.1.3 NONLINEAR OPTIMAL CONTROL FOR MARINE-TURBINE AND SG POWER UNITS

The state-space model of the marine-turbine power generation unit, previously defined in Eq. (2.1), can be written also in the form

$$\dot{x} = f(x) + g_a(x)u_1 + g_b(x)u_2 \tag{2.7}$$

where matrices $f(x) \in R^{5 \times 1}$, $g_a(x) \in R^{5 \times 1}$ and $g_b(x) \in R^{5 \times 1}$ are given by

$$f(x) = \begin{pmatrix} x_2 \\ \frac{1}{J_t}[-K_1(x_1 - x_3) - Dx_2] \\ x_4 \\ \frac{1}{J_g}[K_1(x_1 - x_3)\frac{n_g}{n_t} - D_g x_4 - \omega_0 \frac{V_s x_5 \sin(x_3)}{x_{d\Sigma}'}] \\ -\frac{1}{T_d'}x_5 + \frac{1}{T_{do}}\frac{x_d - x_d'}{x_{d\Sigma}'}V_s \cos(x_3) \end{pmatrix} \quad g_a(x) = \begin{pmatrix} 0 \\ \frac{T_m}{J_t} \\ 0 \\ 0 \\ 0 \end{pmatrix} \quad g_b(x) = \begin{pmatrix} 0 \\ 0 \\ 0 \\ 0 \\ \frac{1}{T_{do}} \end{pmatrix} \tag{2.8}$$

The system undergoes approximate linearization with Taylor series expansion, at each sampling instant, around the temporary operating point (x^*, u^*) where x^* is the present value of the system's state vector and u^* is the last sampled value of the control inputs vector. The linearized model of the system is given by

$$\dot{x} = Ax + Bu + \tilde{d} \tag{2.9}$$

where A, B are the Jacobian matrices of the state-space model and $\tilde{d}$ is the cumulative disturbance term which comprises (i) modeling errors due to the truncation of higher-order terms in the Taylor series expansion, (ii) exogenous perturbations and (iii) sensors measurement noise of any distribution. The Jacobian matrices A and B are given by

$$A = \nabla_x[f(x) + g(x)]\,|_{(x^*, u^*)} \Rightarrow A = \nabla_x f(x)\,|_{(x^*, u^*)} \tag{2.10}$$

$$B = \nabla_u[f(x) + g(x)]\,|_{(x^*, u^*)} \Rightarrow B = g(x)\,|_{(x^*, u^*)} \tag{2.11}$$

The elements of the Jacobian matrix $A = \nabla_x f(x)\,|_{(x^*, u^*)}$ are computed as follows:

First row of the Jacobian matrix $A = \nabla_x f(x)\,|_{(x^*, u^*)}$: $\frac{\partial f_1}{\partial x_1} = 0$, $\frac{\partial f_1}{\partial x_2} = 0$, $\frac{\partial f_1}{\partial x_3} = 0$, $\frac{\partial f_1}{\partial x_4} = 0$, $\frac{\partial f_1}{\partial x_5} = 0$.

Second row of the Jacobian matrix $A = \nabla_x f(x)\,|_{(x^*, u^*)}$: $\frac{\partial f_2}{\partial x_1} = -\frac{K_1}{J_t}$, $\frac{\partial f_2}{\partial x_2} = -\frac{D}{J_t}$, $\frac{\partial f_2}{\partial x_3} = \frac{K_1}{J_t}$, $\frac{\partial f_2}{\partial x_4} = 0$, $\frac{\partial f_2}{\partial x_5} = 0$.

Third row of the Jacobian matrix $A = \nabla_x f(x)\,|_{(x^*, u^*)}$: $\frac{\partial f_3}{\partial x_1} = 0$, $\frac{\partial f_3}{\partial x_2} = 0$, $\frac{\partial f_3}{\partial x_3} = 0$, $\frac{\partial f_3}{\partial x_4} = 1$, $\frac{\partial f_3}{\partial x_5} = 0$.

Fourth row of the Jacobian matrix $A = \nabla_x f(x)\,|_{(x^*, u^*)}$: $\frac{\partial f_4}{\partial x_1} = \frac{K_1}{J_g}\frac{n_g}{n_t}$, $\frac{\partial f_4}{\partial x_2} = 0$, $\frac{\partial f_4}{\partial x_3} = -\frac{K_1}{J_g}\frac{n_g}{n_t} - \frac{\omega_0}{J_g}\frac{V_s x_5 cos(x_3)}{x'_{d\Sigma}}$, $\frac{\partial f_4}{\partial x_4} = -\frac{D_g}{J_g}$, $\frac{\partial f_4}{\partial x_5} = -\frac{\omega_0}{J_g}\frac{V_s sin(x_3)}{x'_{d\Sigma}}$.

Fifth row of the Jacobian matrix $A = \nabla_x f(x)\,|_{(x^*, u^*)}$: $\frac{\partial f_5}{\partial x_1} = 0$, $\frac{\partial f_5}{\partial x_2} = 0$, $\frac{\partial f_5}{\partial x_3} = -\frac{1}{T_{do}}\frac{x_d - x'_d}{x'_{d\Sigma}}V_s sin(x_3)$, $\frac{\partial f_5}{\partial x_4} = 0$, $\frac{\partial f_5}{\partial x_5} = -\frac{1}{T'_d}$.

After linearization around its current operating point, the dynamic model of the marine-turbine and synchronous generator power unit is written as [202]

$$\dot{x} = Ax + Bu + d_1 \tag{2.12}$$

Parameter d_1 stands for the linearization error in the marine turbine and synchronous generator's dynamic model appearing above in Eq. (2.12). The reference setpoints for the state vector of the marine power unit are denoted by $x_d = [x_1^d, \cdots, x_5^d]$. Tracking of this trajectory is achieved after applying the control input u_d. At every time instant, the control input u_d is assumed to differ from the control input u appearing in Eq. (2.12) by an amount equal to Δu, that is $u_d = u + \Delta u$

$$\dot{x}_d = Ax_d + Bu_d + d_2 \tag{2.13}$$

The dynamics of the controlled system described in Eq. (2.12) can be also written as

$$\dot{x} = Ax + Bu + Bu_d - Bu_d + d_1 \tag{2.14}$$

and by denoting $d_3 = -Bu_d + d_1$ as an aggregate disturbance term, one obtains

$$\dot{x} = Ax + Bu + Bu_d + d_3 \tag{2.15}$$

By subtracting Eq. (2.13) from Eq. (2.15) one has

$$\dot{x} - \dot{x}_d = A(x - x_d) + Bu + d_3 - d_2 \tag{2.16}$$

By denoting the tracking error as $e = x - x_d$ and the aggregate disturbance term as $L\tilde{d} = d_3 - d_2$, the tracking error dynamics becomes

$$\dot{e} = Ae + Bu + L\tilde{d} \tag{2.17}$$

where L is a disturbance inputs gain matrix. For the approximately linearized model of the system a stabilizing feedback controller is developed. The controller has the form

$$u(t) = -Ke(t) \tag{2.18}$$

with $K = \frac{1}{r}B^T P$ where P is a positive definite symmetric matrix, which is obtained from the solution of the Riccati equation [202]

$$A^T P + PA + Q - P(\frac{2}{r}BB^T - \frac{1}{\rho^2}LL^T)P = 0 \tag{2.19}$$

where Q is a positive semi-definite symmetric matrix. Through Lyapunov stability analysis it will be shown that the proposed nonlinear control scheme assures H_∞ tracking performance for the marine-turbine and synchronous generator power unit, and that under moderate conditions asymptotic convergence to the reference setpoints (global stability) is achieved [198]. As shown before, the tracking error dynamics for the marine-turbine and synchronous generator power unit is written in the form [198]

$$\dot{e} = Ae + Bu + L\tilde{d} \tag{2.20}$$

where in the marine-turbine and synchronous generator power unit's case $L \in R^{5 \times 5}$ is the disturbance inputs matrix. Variable $\tilde{d}$ denotes model uncertainties and external disturbances of the marine-turbine and synchronous generator's dynamic model. The following Lyapunov equation is considered

$$V = \frac{1}{2}e^T Pe \tag{2.21}$$

where $e = x - x_d$ is the tracking error. By differentiating with respect to time, one obtains

$$\dot{V} = \frac{1}{2}\dot{e}^T Pe + \frac{1}{2}e^T P\dot{e} \Rightarrow$$
$$\dot{V} = \frac{1}{2}[Ae + Bu + L\tilde{d}]^T P + \frac{1}{2}e^T P[Ae + Bu + L\tilde{d}] \Rightarrow \tag{2.22}$$

$$\dot{V} = \frac{1}{2}[e^T A^T + u^T B^T + \tilde{d}^T L^T]Pe +$$
$$+\frac{1}{2}e^T P[Ae + Bu + L\tilde{d}] \Rightarrow \tag{2.23}$$

$$\dot{V} = \frac{1}{2}e^T A^T Pe + \frac{1}{2}u^T B^T Pe + \frac{1}{2}\tilde{d}^T L^T Pe +$$
$$\frac{1}{2}e^T PAe + \frac{1}{2}e^T PBu + \frac{1}{2}e^T PL\tilde{d} \tag{2.24}$$

The previous equation is rewritten as

$$\dot{V} = \frac{1}{2}e^T (A^T P + PA)e + (\frac{1}{2}u^T B^T Pe + \frac{1}{2}e^T PBu) +$$
$$+(\frac{1}{2}\tilde{d}^T L^T Pe + \frac{1}{2}e^T PL\tilde{d}) \tag{2.25}$$

Assumption: For given positive definite matrix Q and coefficients r and ρ there exists a positive definite matrix P, which is the solution of the following matrix equation

$$A^T P + PA = -Q + P(\frac{2}{r}BB^T - \frac{1}{\rho^2}LL^T)P \tag{2.26}$$

Moreover, the following feedback control law is applied to the system

$$u = -\frac{1}{r}B^T Pe \tag{2.27}$$

By substituting Eq. (2.26) and Eq. (2.27), one obtains

$$\dot{V} = \frac{1}{2}e^T[-Q + P(\frac{2}{r}BB^T - \frac{1}{2\rho^2}LL^T)P]e +$$
$$+e^T PB(-\frac{1}{r}B^T Pe) + e^T PL\tilde{d} \Rightarrow \tag{2.28}$$

$$\dot{V} = -\frac{1}{2}e^T Qe + (\frac{2}{r}PBB^T Pe - \frac{1}{2\rho^2}e^T PLL^T Pe$$
$$-\frac{1}{r}e^T PBB^T Pe) + e^T PL\tilde{d} \tag{2.29}$$

which after intermediate operations gives

$$\dot{V} = -\frac{1}{2}e^T Qe - \frac{1}{2\rho^2}e^T PLL^T Pe + e^T PL\tilde{d} \tag{2.30}$$

or, equivalently

$$\dot{V} = -\frac{1}{2}e^T Qe - \frac{1}{2\rho^2}e^T PLL^T Pe +$$
$$+\frac{1}{2}e^T PL\tilde{d} + \frac{1}{2}\tilde{d}^T L^T Pe \tag{2.31}$$

Lemma: The following inequality holds

$$\frac{1}{2}e^T PL\tilde{d} + \frac{1}{2}\tilde{d}L^T Pe - \frac{1}{2\rho^2}e^T PLL^T Pe \leq \frac{1}{2}\rho^2\tilde{d}^T\tilde{d} \tag{2.32}$$

Proof: The binomial $(\rho\alpha - \frac{1}{\rho}b)^2$ is considered. Expanding the left part of the above inequality one gets

$$\rho^2a^2 + \frac{1}{\rho^2}b^2 - 2ab \geq 0 \Rightarrow \frac{1}{2}\rho^2a^2 + \frac{1}{2\rho^2}b^2 - ab \geq 0 \Rightarrow$$
$$ab - \frac{1}{2\rho^2}b^2 \leq \frac{1}{2}\rho^2a^2 \Rightarrow \frac{1}{2}ab + \frac{1}{2}ab - \frac{1}{2\rho^2}b^2 \leq \frac{1}{2}\rho^2a^2 \tag{2.33}$$

The following substitutions are carried out: $a = \tilde{d}$ and $b = e^T PL$ and the previous relation becomes

$$\frac{1}{2}\tilde{d}^T L^T Pe + \frac{1}{2}e^T PL\tilde{d} - \frac{1}{2\rho^2}e^T PLL^T Pe \leq \frac{1}{2}\rho^2\tilde{d}^T\tilde{d} \tag{2.34}$$

Eq. (2.34) is substituted in Eq. (2.31) and the inequality is enforced, thus giving

$$\dot{V} \leq -\frac{1}{2}e^T Qe + \frac{1}{2}\rho^2\tilde{d}^T\tilde{d} \tag{2.35}$$

Eq. (2.35) shows that the H_∞ tracking performance criterion is satisfied. The integration of $\dot{V}$ from 0 to T gives

$$\int_0^T \dot{V}(t)dt \leq -\frac{1}{2}\int_0^T ||e||_Q^2 dt + \frac{1}{2}\rho^2\int_0^T ||\tilde{d}||^2 dt \Rightarrow$$
$$2V(T) + \int_0^T ||e||_Q^2 dt \leq 2V(0) + \rho^2\int_0^T ||\tilde{d}||^2 dt \tag{2.36}$$

Moreover, if there exists a positive constant $M_d > 0$ such that

$$\int_0^\infty ||\tilde{d}||^2 dt \leq M_d \tag{2.37}$$

then one gets

$$\int_0^\infty ||e||_Q^2 dt \leq 2V(0) + \rho^2 M_d \tag{2.38}$$

Thus, the integral $\int_0^\infty ||e||_Q^2 dt$ is bounded. Moreover, $V(T)$ is bounded and from the definition of the Lyapunov function V in Eq. (2.21) it becomes clear that $e(t)$ will be also bounded since $e(t) \in \Omega_e = \{e | e^T Pe \leq 2V(0) + \rho^2 M_d\}$. According to the above and with the use of Barbalat's Lemma, one obtains $lim_{t\to\infty}e(t) = 0$.

By following the stages of the stability proof one arrives at Eq. (2.35) which shows that the H-infinity tracking performance criterion holds. By selecting the attenuation coefficient ρ to be sufficiently small and in particular to satisfy $\rho^2 < ||e||_Q^2/||\tilde{d}||^2$ one has that the first derivative of the Lyapunov function is upper bounded by 0. This condition holds at each sampling instant and consequently global stability for the control loop can be concluded.

2.1.4 LIE ALGEBRA-BASED CONTROL FOR MARINE-TURBINE AND SG POWER UNITS

For the state-space model of the marine-turbine and synchronous generator power unit that was given in Eq. (2.1) the following linearizing outputs of the system are defined:

$$y_1 = h_1(x) = x_1 \quad y_2 = h_2(x) = x_3 \tag{2.39}$$

For this state-space model the following Lie derivatives are computed:

$$L_f h_1 = \frac{\partial h_1}{\partial x_1} f_1 + \frac{\partial h_1}{\partial x_2} f_2 + \frac{\partial h_1}{\partial x_3} f_3 + \frac{\partial h_1}{\partial x_4} f_4 + \frac{\partial h_1}{\partial x_5} f_5$$
$$\Rightarrow L_f h_1 = x_2 \tag{2.40}$$

$$L_{g_a} h_1 = \frac{\partial h_1}{\partial x_1} g_{a1} + \frac{\partial h_1}{\partial x_2} g_{a2} + \frac{\partial h_1}{\partial x_3} g_{a3} + \frac{\partial h_1}{\partial x_4} g_{a4} + \frac{\partial h_1}{\partial x_5} g_{a5}$$
$$\Rightarrow L_{g_a} h_1 = 0 \tag{2.41}$$

$$L_{g_b} h_1 = \frac{\partial h_1}{\partial x_1} g_{b1} + \frac{\partial h_1}{\partial x_2} g_{b2} + \frac{\partial h_1}{\partial x_3} g_{b3} + \frac{\partial h_1}{\partial x_4} g_{b4} + \frac{\partial h_1}{\partial x_5} g_{b5}$$
$$\Rightarrow L_{g_b} h_1 = 0 \tag{2.42}$$

Equivalently, using that $L_f h_1 = x_2$, it holds that

$$L_f^2 h_1 = \frac{\partial x_2}{\partial x_1} f_1 + \frac{\partial x_2}{\partial x_2} f_2 + \frac{\partial x_2}{\partial x_3} f_3 + \frac{\partial x_2}{\partial x_4} f_4 + \frac{\partial x_2}{\partial x_5} f_5$$
$$\Rightarrow L_f^2 h_1 = f_2 \tag{2.43}$$

$$L_{g_a} L_f h_1 = \frac{\partial x_2}{\partial x_1} g_{a1} + \frac{\partial x_2}{\partial x_2} g_{a2} + \frac{\partial x_2}{\partial x_3} g_{a3} + \frac{\partial x_2}{\partial x_4} g_{a4} + \frac{\partial x_2}{\partial x_5} g_{a5}$$
$$\Rightarrow L_{g_a} L_f h_1 = \frac{T_m}{J_t} \tag{2.44}$$

$$L_{g_b} L_f h_1 = \frac{\partial x_2}{\partial x_1} g_{b1} + \frac{\partial x_2}{\partial x_2} g_{b2} + \frac{\partial x_2}{\partial x_3} g_{b3} + \frac{\partial x_2}{\partial x_4} g_{b4} + \frac{\partial x_2}{\partial x_5} g_{b5}$$
$$\Rightarrow L_{g_b} L_f h_1 = 0 \tag{2.45}$$

By denoting $z_1 = h_1$, it holds that

$$\dot{z}_1 = \dot{x}_1 \Rightarrow \dot{z}_1 = x_2 \Rightarrow \dot{z}_1 = L_f h_1 \Rightarrow$$
$$\ddot{z}_1 = \dot{x}_2 \Rightarrow \ddot{z}_1 = f_2(x) + g_{a_2}(x)u_1 + g_{b_2}(x)u_2 \Rightarrow \tag{2.46}$$
$$\ddot{z}_1 = L_f^2 h_1 + L_{g_a} L_f h_1 u_1 + L_{g_b} L_f h_1 u_2$$

In a similar manner, one obtains

$$L_f h_2 = \frac{\partial h_2}{\partial x_1} f_1 + \frac{\partial h_2}{\partial x_2} f_2 + \frac{\partial h_2}{\partial x_3} f_3 + \frac{\partial h_2}{\partial x_4} f_4 + \frac{\partial h_2}{\partial x_5} f_5$$
$$\Rightarrow L_f h_2 = x_4 \tag{2.47}$$

$$L_{g_a} h_2 = \frac{\partial h_2}{\partial x_1} g_{a1} + \frac{\partial h_2}{\partial x_2} g_{a2} + \frac{\partial h_2}{\partial x_3} g_{a3} + \frac{\partial h_2}{\partial x_4} g_{a4} + \frac{\partial h_2}{\partial x_5} g_{a5}$$
$$\Rightarrow L_{g_a} h_2 = 0 \tag{2.48}$$

$$L_{g_b} h_2 = \frac{\partial h_2}{\partial x_1} g_{b1} + \frac{\partial h_2}{\partial x_2} g_{b2} + \frac{\partial h_2}{\partial x_3} g_{b3} + \frac{\partial h_2}{\partial x_4} g_{b4} + \frac{\partial h_2}{\partial x_5} g_{b5}$$
$$\Rightarrow L_{g_b} h_2 = 0 \tag{2.49}$$

Equivalently, by denoting $L_f h_2 = x_4$, one has

$$L_f^2 h_2 = \frac{\partial x_4}{\partial x_1} f_1 + \frac{\partial x_4}{\partial x_2} f_2 + \frac{\partial x_4}{\partial x_3} f_3 + \frac{\partial x_4}{\partial x_4} f_4 + \frac{\partial x_4}{\partial x_5} f_5$$
$$\Rightarrow L_f^2 h_2 = f_4 \tag{2.50}$$

$$L_{g_a} L_f h_2 = \frac{\partial f_4}{\partial x_1} g_{a1} + \frac{\partial f_4}{\partial x_2} g_{a2} + \frac{\partial f_4}{\partial x_3} g_{a3} + \frac{\partial f_4}{\partial x_4} g_{a4} + \frac{\partial f_4}{\partial x_5} g_{a5}$$
$$\Rightarrow L_{g_a} L_f h_2 = 0 \tag{2.51}$$

$$L_{g_b} L_f h_2 = \frac{\partial f_4}{\partial x_1} g_{b1} + \frac{\partial f_4}{\partial x_2} g_{b2} + \frac{\partial f_4}{\partial x_3} g_{b3} + \frac{\partial f_4}{\partial x_4} g_{b4} + \frac{\partial f_4}{\partial x_5} g_{b5}$$
$$\Rightarrow L_{g_b} L_f h_2 = 0 \tag{2.52}$$

Additionally, using that $L_f^2 h_2 = f_4$ one has

$$L_f^3 h_2 = \frac{\partial f_4}{\partial x_1} f_1 + \frac{\partial f_4}{\partial x_2} f_2 + \frac{\partial f_4}{\partial x_3} f_3 + \frac{\partial f_4}{\partial x_4} f_4 + \frac{\partial f_4}{\partial x_5} f_5$$
$$\Rightarrow L_f^3 h_2 = \frac{K_1}{J_g} \frac{n_g}{n_t} f_1 - [\frac{K_1}{J_g} \frac{n_g}{n_t} - \frac{\omega_0}{J_g} \frac{V_s x_5 cos(x_3)}{x'_{d\Sigma}}] f_3 - \frac{D_g}{J_g} f_4 - [\frac{\omega_0}{J_g} \frac{V_s cos(x_3)}{x'_{d\Sigma}}] f_5 \tag{2.53}$$

$$L_{g_a} L_f^2 h_2 = [\frac{K_1}{J_g} \frac{n_g}{n_t}] g_{a1} + [\frac{K_1}{J_g} \frac{n_g}{n_t} - \frac{\omega_0}{J_g} \frac{V_s x_5 cos(x_3)}{x'_{d\Sigma}}] g_{a3} + [-\frac{D_g}{J_g}] g_{a4} + [-\frac{\omega_0}{J_g} \frac{V_s sin(x_3)}{x'_{d\Sigma}}] g_{a5}$$
$$\Rightarrow L_{g_a} L_f^2 h_2 = 0 \tag{2.54}$$

$$L_{g_b} L_f^2 h_2 = [\frac{K_1}{J_g} \frac{n_g}{n_t}] g_{b1} + [\frac{K_1}{J_g} \frac{n_g}{n_t} - \frac{\omega_0}{J_g} \frac{V_s x_5 cos(x_3)}{x'_{d\Sigma}}] g_{b3} + [-\frac{D_g}{J_g}] g_{b4} + [-\frac{\omega_0}{J_g} \frac{V_s sin(x_3)}{x'_{d\Sigma}}] g_{b5}$$
$$\Rightarrow L_{g_b} L_f^2 h_2 = [-\frac{\omega_0}{J_g} \frac{V_s sin(x_3)}{x'_{d\Sigma}}] \frac{1}{T_{do}} \tag{2.55}$$

One defines $z_2 = h_2$, thus it holds

$$\dot z_2 = \dot x_3 \Rightarrow \dot z_2 = x_4 \Rightarrow \dot z_2 = L_f h_2 \Rightarrow$$
$$\ddot z_2 = \dot x_4 \Rightarrow \ddot z_2 = f_4(x) \Rightarrow \ddot z_2 = L_f^2 h_2 \Rightarrow$$
$$z_2^{(3)} = \dot f_4 \Rightarrow z_2^3 = \frac{\partial f_4}{\partial x_1} \dot x_1 + \frac{\partial f_4}{\partial x_2} \dot x_2 + \frac{\partial f_4}{\partial x_3} \dot x_3 + \frac{\partial f_4}{\partial x_4} \dot x_4 + \frac{\partial f_4}{\partial x_5} \dot x_5 \tag{2.56}$$

which gives

$$z_2^{(3)} = [\frac{K_1}{J_g} \frac{n_g}{n_t}] f_1 - [\frac{K_1}{J_g} \frac{n_g}{n_t} - \frac{\omega_0}{J_g} \frac{V_s x_5 cos(x_3)}{x'_{d\Sigma}}] f_3 - \frac{D_g}{J_g} f_4 - [-\frac{\omega_0}{J_g} \frac{V_s sin(x_3)}{x'_{d\Sigma}}] f_5 -$$
$$[-\frac{\omega_0}{J_g} \frac{V_s sin(x_3)}{x'_{d\Sigma}}] \frac{1}{T_{do}} u_2 \tag{2.57}$$

or equivalently

$$z_2^{(3)} = L_f^3 h_2 + L_{g_a} L_f^2 h_2 u_1 + L_{g_b} L_f^2 h_2 u_2 \tag{2.58}$$

Consequently, one arrives at the following linearized form for the marine-turbine and synchronous generator power unit

$$\ddot{z}_1 = L_f^2 h_1 + L_{g_a} L_f h_1 u_1 + L_{g_b} L_f h_1 u_2$$
$$z_2^{(3)} = L_f^3 h_2 + L_{g_a} L_f^2 h_2 u_1 + L_{g_b} L_f^2 h_2 u_2 \qquad (2.59)$$

Besides, one defines the following virtual control inputs

$$v_1 = L_f^2 h_1 + L_{g_a} L_f h_1 u_1 + L_{g_b} L_f h_1 u_2$$
$$v_2 = L_f^3 h_2 + L_{g_a} L_f^2 h_2 u_1 + L_{g_b} L_f^2 h_2 u_2 \qquad (2.60)$$

Thus. the linearized dynamics of the system is written as

$$\ddot{z}_1 = v_1 \qquad z_2^{(3)} = v_2 \qquad (2.61)$$

or, by defining the state vector $Z = [z_1, \dot{z}_1, z_2, \dot{z}_2, \ddot{z}_2]^T$ in matrix form one has

$$
\begin{pmatrix} \dot{z}_1 \\ \ddot{z}_1 \\ \dot{z}_2 \\ \ddot{z}_2 \\ z_2^{(3)} \end{pmatrix} =
\begin{pmatrix} 0 & 1 & 0 & 0 & 0 \\ 0 & 0 & 0 & 0 & 0 \\ 0 & 0 & 0 & 1 & 0 \\ 0 & 0 & 0 & 0 & 1 \\ 0 & 0 & 0 & 0 & 0 \end{pmatrix}
\begin{pmatrix} z_1 \\ \dot{z}_1 \\ z_2 \\ \dot{z}_2 \\ \ddot{z}_2 \end{pmatrix} +
\begin{pmatrix} 0 & 0 \\ 1 & 0 \\ 0 & 0 \\ 0 & 0 \\ 0 & 1 \end{pmatrix}
\begin{pmatrix} v_1 \\ v_2 \end{pmatrix} \qquad (2.62)
$$

The associated state-observer for the linearized system is

$$\dot{\hat{Z}} = AZ + Bv + K(Z_m - \hat{Z}_m) \qquad (2.63)$$

The gain K of the state observer for the linearized system can be computed using the linear Kalman Filter algorithm. Then by defining the measurements vector to be $Z_m = h(x) = [h_1(x) \ h_2(x)]^T$, the state observer for the initial nonlinear system is

$$\dot{\hat{x}} = f(\hat{x}) + g_a(\hat{x})u_1 + g_b(\hat{x})u_2 + L(h(x) - h(\hat{x})) \qquad (2.64)$$

where the gain of the state observer for the nonlinear system is

$$L = (J_\phi(\hat{x}))^{-1} K \qquad (2.65)$$

with the Jacobian matrix $J_\phi(\hat{x})$ to be given by

$$
J_\phi(\hat{x}) =
\begin{pmatrix}
\frac{\partial h_1}{\partial x_1} & \frac{\partial h_1}{\partial x_2} & \frac{\partial h_1}{\partial x_3} & \frac{\partial h_1}{\partial x_4} & \frac{\partial h_1}{\partial x_5} \\[6pt]
\frac{\partial L_f h_1}{\partial x_1} & \frac{\partial L_f h_1}{\partial x_2} & \frac{\partial L_f h_1}{\partial x_3} & \frac{\partial L_f h_1}{\partial x_4} & \frac{\partial L_f h_1}{\partial x_5} \\[6pt]
\frac{\partial h_2}{\partial x_1} & \frac{\partial h_2}{\partial x_2} & \frac{\partial h_2}{\partial x_3} & \frac{\partial h_2}{\partial x_4} & \frac{\partial h_2}{\partial x_5} \\[6pt]
\frac{\partial L_f h_2}{\partial x_1} & \frac{\partial L_f h_2}{\partial x_2} & \frac{\partial L_f h_2}{\partial x_3} & \frac{\partial L_f h_2}{\partial x_4} & \frac{\partial L_f h_2}{\partial x_5} \\[6pt]
\frac{\partial L_f^2 h_2}{\partial x_1} & \frac{\partial L_f^2 h_2}{\partial x_2} & \frac{\partial L_f^2 h_2}{\partial x_3} & \frac{\partial L_f^2 h_2}{\partial x_4} & \frac{\partial L_f^2 h_2}{\partial x_5}
\end{pmatrix} \qquad (2.66)
$$

where $h_1 = x_1$ thus: $\frac{\partial h_1}{\partial x_1} = 1$, $\frac{\partial h_1}{\partial x_2} = 0$, $\frac{\partial h_1}{\partial x_3} = 0$, $\frac{\partial h_1}{\partial x_4} = 0$, $\frac{\partial h_1}{\partial x_5} = 0$.

$L_f h_1 = x_2$ thus: $\frac{\partial L_f h_1}{\partial x_1} = 0$, $\frac{\partial L_f h_1}{\partial x_2} = 1$, $\frac{\partial L_f h_1}{\partial x_3} = 0$, $\frac{\partial L_f h_1}{\partial x_4} = 0$, $\frac{\partial L_f h_1}{\partial x_5} = 0$

and $h_2 = x_3$ thus: $\frac{\partial h_2}{\partial x_1} = 0$, $\frac{\partial h_2}{\partial x_2} = 0$, $\frac{\partial h_2}{\partial x_3} = 1$, $\frac{\partial h_2}{\partial x_4} = 0$ and $\frac{\partial h_2}{\partial x_5} = 0$

$L_f h_2 = x_4$ thus: $\frac{\partial L_f h_2}{\partial x_1} = 0$, $\frac{\partial L_f h_2}{\partial x_2} = 0$, $\frac{\partial L_f h_2}{\partial x_3} = 0$, $\frac{\partial L_f h_2}{\partial x_4} = 1$ and $\frac{\partial L_f h_2}{\partial x_5} = 0$.

$L_f^2 h_2 = f_4$ thus: $\frac{\partial L_f^2 h_2}{\partial x_1} = \frac{K_1}{J_g} \frac{n_g}{n_t}$, $\frac{\partial L_f^2 h_2}{\partial x_2} = 0$, $\frac{\partial L_f^2 h_2}{\partial x_3} = -\frac{K_1}{J_g} \frac{n_g}{n_t} - \frac{\omega_0}{J_g} \frac{V_s x_5 cos(x_3)}{x_{d\Sigma}'}]$, $\frac{\partial L_f^2 h_2}{\partial x_4} = -\frac{D_g}{J_g}$ and $\frac{\partial L_f^2 h_2}{\partial x_5} = -\frac{\omega_0}{J_g} \frac{V_s cos(x_3)}{x_{d\Sigma}'}]$.

2.1.5   FLATNESS-BASED CONTROL FOR MARINE-TURBINE AND SG POWER UNITS

The flat outputs of the marine power generation unit, with the state-space model of Eq. (2.1), are: $y_1 = x_1$ and $y_3 = x_3$. From the first row of the states-space model one has

$$x_2 = \dot{x}_1 \tag{2.67}$$

Consequently, x_2 is a differential function of the flat outputs of the system. From the third row of the state-space model, one obtains

$$x_4 = \dot{x}_2 \tag{2.68}$$

Consequently, x_4 is a differential function of the flat outputs of the system. From the 4th row of the state-space model one has

$$\dot{x}_1 - \frac{1}{J_t}[K_1(x_1 - x_3)\frac{n_g}{n_t} + D_g x_4] = -\frac{\frac{\omega_0}{J_g} V_s x_5 sin(x_3)}{x_{d\Sigma}'} \Rightarrow$$
$$x_5 = \frac{\{\dot{x}_1 - \frac{1}{J_t}[K_1(x_1-x_3)\frac{n_g}{n_t} + D_g x_4]\}}{-\frac{\omega_0}{J_g} \frac{V_s sin(x_3)}{x_{d\Sigma}'}} \tag{2.69}$$

Consequently, x_5 is a differential function of the flat outputs of the system. Besides, from the second row of the state-space model one has

$$u_1 = \frac{J_m}{T_m}\{\dot{x}_2 - \frac{1}{J_t}[-K_1(x_1 - x_3) - Dx_2]\} \tag{2.70}$$

Consequently, u_1 is a differential function of the system's flat outputs. Additionally, from the first row of the state-space model, one obtains that control input u_2 is also a differential function of the system's flat outputs

$$u_2 = T_{do}\{\dot{x}_5 + \frac{1}{T_d'}x_5 - \frac{1}{T_{do}} \frac{x_d - x_d'}{x_{d\Sigma}'} V_s cos(x_3)\} \tag{2.71}$$

As a result of the above, all state variables and the control inputs of the marine-turbine and synchronous generator power system can be written as differential functions of the flat outputs of the system. Consequently, the system is differential flat and is also input-output linearizable [195].

Indeed, by differentiating twice the first flat output of the system, one obtains

$$\ddot{x}_1 = \frac{1}{J_t}[-K_1(x_1 - x_3) - Dx_2] + \frac{T_m}{J_t}u_1 \tag{2.72}$$

Besides, by differentiating twice the second flat output of the system, one obtains

$$\ddot{x}_3 = \frac{1}{J_g}[K_1(x_1 - x_3)\frac{n_g}{n_t} - D_g x_4 - \omega_0 \frac{V_s x_5 sin(x_3)}{x'_{d\Sigma}}] \tag{2.73}$$

By differentiating x_3 once more with respect to time, one obtains

$$x_3^{(3)} = \frac{1}{J_g}[K_1(\dot{x}_1 - \dot{x}_3)\frac{n_g}{n_t} - D_g\dot{x}_4 - \omega_0\frac{V_s\dot{x}_5 sin(x_3)}{x'_{d\Sigma}} - \omega_0\frac{V_s x_5 cos(x_3)\dot{x}_3}{x'_{d\Sigma}}] \tag{2.74}$$

By substituting the time derivatives $\dot{x}_1$, $\dot{x}_3$, $\dot{x}_4$ and $\dot{x}_5$ in the previous relation, one obtains

$$\begin{aligned} x_3^{(3)} &= \frac{1}{J_g}K_1(x_2 - x_4)\frac{n_g}{n_t} \\ &- \frac{D_g}{J_g}\{\frac{1}{J_g}[K_1(x_1 - x_3)\frac{n_g}{n_t} - D_g x_4 - \omega_0\frac{V_s x_5 sin(x_3)}{x'_{d\Sigma}}]\} - \\ &- \omega_0\frac{V_s sin(x_3)}{x'_{d\Sigma}}\{-\frac{1}{T'_d}x_5 + \frac{1}{T_{do}}\frac{x_d - x'_d}{x'_{d\Sigma}}V_s cos(x_3) + \frac{1}{T_{do}}u_2\} - \omega_0\frac{V_s x_5 cos(x_3)x_4}{x'_{d\Sigma}} \end{aligned} \tag{2.75}$$

Consequently, the marine turbine and synchronous generator power unit is written in the fllowing input-output linearized form

$$\ddot{x}_1 = f_1(x) + g_{11}(x)u_1 + g_{21}(x)u_2 \tag{2.76}$$

where about functions $f_1(x)(x)$, $g_{11}(x)$, $g_{21}(x)$ one has

$$f_1(x) = \frac{1}{J_t}[-K_1(x_1 - x_3) - Dx_2] \tag{2.77}$$

$$g_{11}(x) = \frac{T_m}{J_t} \quad g_{21}(x) = 0 \tag{2.78}$$

$$x_3^{(3)} = f_2(x) + g_{12}(x)u_1 + g_{22}(x)u_2 \tag{2.79}$$

where about functions $f_2(x)(x)$, $g_{12}(x)$, $g_{22}(x)$ one has

$$\begin{aligned} f_2(x) &= \frac{1}{J_g}K_1(x_2 - x_4)\frac{n_g}{n_t} \\ &- \frac{D_g}{J_g}\{\frac{1}{J_g}[K_1(x_1 - x_3)\frac{n_g}{n_t} - D_g x_4 - \omega_0\frac{V_s x_5 sin(x_3)}{x'_{d\Sigma}}]\} - \\ &- \omega_0\frac{V_s sin(x_3)}{x'_{d\Sigma}}\{-\frac{1}{T'_d}x_5 + \frac{1}{T_{do}}\frac{x_d - x'_d}{x'_{d\Sigma}}V_s cos(x_3)\} - \omega_0\frac{V_s x_5 cos(x_3)x_4}{x'_{d\Sigma}} \end{aligned} \tag{2.80}$$

$$g_{12}(x) = 0 \quad g_{22}(x) = -\omega_0 \frac{V_s sin(x_3)}{x_{d\Sigma}'} \frac{1}{T_{do}}$$

(2.81)

Thus, one arrives at the following input-output linearized form for the system of the marine turbine and synchronous generator

$$\begin{pmatrix} \ddot{x}_1 \\ x_3^{(3)} \end{pmatrix} = \begin{pmatrix} f_1(x) \\ f_2(x) \end{pmatrix} + \begin{pmatrix} g_{11}(x) & 0 \\ 0 & g_{22}(x) \end{pmatrix} \begin{pmatrix} u_1 \\ u_2 \end{pmatrix}$$

(2.82)

Next, a stabilizing feedback controller for the marine-turbine and synchronous generator power unit is designed. The input-output linearized description of the power unit is rewritten after including additive disturbance terms in it. Thus, one has

$$\ddot{x}_1 = f_1(x) + g_{11}(xu_1) + g_{21}(x)u_2 + \tilde{d}_1$$

$$x_3^{(3)} = f_2(x) + g_{12}(xu_1) + g_{22}(x)u_2 + \tilde{d}_2$$

(2.83)

The following virtual control inputs are defined: $v_1 = f_1(x) + g_{11}(x)u_1 + g_{21}(x)u_2$ and $v_2 = f_2(x) + g_{12}(xu_1) + g_{22}(x)u_2$. The disturbances $\tilde{d}_1$ and $\tilde{d}_2$ are denoted as additional state variables. Consequently, the state variables of the system become: $z_1 = x_1, z_2 = \dot{x}_1, z_3 = x_2, z_4 = \dot{x}_2, z_5 = \ddot{x}_2, z_6 = \tilde{d}_1, z_7 = \dot{\tilde{d}}_1, z_8 = \tilde{d}_2, z_9 = \dot{\tilde{d}}_2$.

By assuming that the second-order time derivatives of the additive disturbances are given by $\ddot{\tilde{d}}_1 = f_{\tilde{d}_1}$ and $\ddot{\tilde{d}}_2 = f_{\tilde{d}_2}$, the extended state-space description of the system is written as

$$\begin{pmatrix} \dot{z}_1 \\ \dot{z}_2 \\ \dot{z}_3 \\ \dot{z}_4 \\ \dot{z}_5 \\ \dot{z}_6 \\ \dot{z}_7 \\ \dot{z}_8 \\ \dot{z}_9 \end{pmatrix} = \begin{pmatrix} 0 & 1 & 0 & 0 & 0 & 0 & 0 & 0 & 0 \\ 0 & 0 & 0 & 0 & 0 & 1 & 0 & 0 & 0 \\ 0 & 0 & 0 & 1 & 0 & 0 & 0 & 0 & 0 \\ 0 & 0 & 0 & 0 & 1 & 0 & 0 & 0 & 0 \\ 0 & 0 & 0 & 0 & 0 & 0 & 0 & 1 & 0 \\ 0 & 0 & 0 & 0 & 0 & 0 & 1 & 0 & 0 \\ 0 & 0 & 0 & 0 & 0 & 0 & 0 & 0 & 0 \\ 0 & 0 & 0 & 0 & 0 & 0 & 0 & 0 & 1 \\ 0 & 0 & 0 & 0 & 0 & 0 & 0 & 0 & 0 \end{pmatrix} \begin{pmatrix} z_1 \\ z_2 \\ z_3 \\ z_4 \\ z_5 \\ z_6 \\ z_7 \\ z_8 \\ z_9 \end{pmatrix} + \begin{pmatrix} 0 & 0 & 0 & 0 \\ 1 & 0 & 0 & 0 \\ 0 & 0 & 0 & 0 \\ 0 & 0 & 0 & 0 \\ 0 & 1 & 0 & 0 \\ 0 & 0 & 0 & 0 \\ 0 & 0 & 1 & 0 \\ 0 & 0 & 0 & 0 \\ 0 & 0 & 0 & 1 \end{pmatrix} \begin{pmatrix} v_1 \\ v_2 \\ f_{\tilde{d}_1} \\ f_{\tilde{d}_2} \end{pmatrix}$$

(2.84)

Denoting the state vector $Z = [z_1, z_2, z_3, z_4, z_5, z_5, z_6, z_7, z_8, z_9]^T$, the associated outputs of the system are z_1 and z_3, thus the measurement equation of the system becomes

$$\begin{pmatrix} z_1^m \\ z_2^m \end{pmatrix} = \begin{pmatrix} 1 & 0 & 0 & 0 & 0 & 0 & 0 & 0 & 0 \\ 0 & 0 & 1 & 0 & 0 & 0 & 0 & 0 & 0 \end{pmatrix} Z$$

(2.85)

or equivalently the extended state-space description of the power unit is written as

$$\dot{Z} = A_e Z + B_e v$$
$$Z_e^m = C_e Z$$

(2.86)

The stabilizing feedback control for the marine-turbine and synchronous generator power unit is given by

$$
\begin{aligned}
v_1 &= \ddot{z}_{1,d} - k_{11}(\dot{z}_1 - \dot{z}_{1,d}) - k_{21}(z_1 - z_{1,d}) - \hat{d}_1 \\
v_2 &= z_{2,d}^{(3)} - k_{12}(\ddot{z}_3 - \ddot{z}_{3,d}) - k_{22}(\dot{z}_3 - \dot{z}_{3,d}) - k_{32}(z_3 - z_{3,d}) - \hat{d}_2
\end{aligned}
\tag{2.87}
$$

To obtain estimates of the additive disturbance terms $\hat{d}_1$ and $\hat{d}_2$ the following disturbance observer is used

$$
\dot{Z} = A_{e,o}Z + B_{e,o}v + K_f(Z_e^m - \hat{Z}_e^m)
\tag{2.88}
$$

where the matrices which appear in the observer's equation are $A_{e,o} = A_e$, $C_{e,o} = C_e$ and

$$
B_{e,o}^T = \begin{pmatrix} 0 & 1 & 0 & 0 & 0 & 0 & 0 & 0 & 0 \\ 0 & 0 & 0 & 0 & 1 & 0 & 0 & 0 & 0 \end{pmatrix}
\tag{2.89}
$$

while the observer's gain matrix K_f is given by Kalman Filter's recursion [195]. The control inputs which are finally applied to the initial nonlinear model of the marine-turbine power generation unit are

$$
\begin{pmatrix} u_1 \\ u_2 \end{pmatrix} = \begin{pmatrix} g_{11}(x) & g_{21}(x) \\ g_{12}(x) & g_{22}(x) \end{pmatrix}^{-1} \begin{pmatrix} v_1 - f_1(x) \\ v_2 - f_2(x) \end{pmatrix}
\tag{2.90}
$$

2.1.6 FLATNESS-BASED CONTROL IN SUCCESSIVE LOOPS FOR MARINE-TURBINE AND SG POWER UNITS

The state-space model of the marine power unit, that was previously given in Eq. (2.1) is written in a row-per-row form:

$$
\dot{x}_1 = x_2
\tag{2.91}
$$

$$
\dot{x}_2 = \frac{1}{J_t}[-K_1(x_1 - x_3) - Dx_2] + \frac{T_m}{J_t}u_1
\tag{2.92}
$$

$$
\dot{x}_3 = x_4
\tag{2.93}
$$

$$
\dot{x}_4 = \frac{1}{J_g}[K_1(x_1 - x_3)\frac{n_g}{n_t} - D_g x_4 - \omega_0 \frac{V_s x_5 \sin(x_3)}{x'_{d\Sigma}}]
\tag{2.94}
$$

$$
\dot{x}_5 = -\frac{1}{T'_d}x_5 + \frac{1}{T_{do}}\frac{x_d - x'_d}{x'_{d\Sigma}}V_s\cos(x_3) + \frac{1}{T_{do}}u_2
\tag{2.95}
$$

Next, using the notation: (i) for Eq. (2.91) $f_1(x) = 0$, $g_1(x) = 1$, (ii) for Eq. (2.92) $f_2(x) = \frac{1}{J_t}[-K_1(x_1 - x_3) - Dx_2]$, $g_2(x) = \frac{T_m}{J_t}$, (iii) for Eq. (2.93) $f_3(x) = 0$, $g_3(x) = 1$,

(iv) for Eq. (2.94) $f_4(x) = \frac{1}{J_g}[K_1(x_1 - x_3)\frac{n_g}{n_t} - D_g x_4 - \omega_0 \frac{V_s x_5 \sin(x_3)}{x_{d_\Sigma}'}]$, $g_4(x) = 0$ and

(v) for Eq. (2.95) $f_5(x) = -\frac{1}{T_d'} x_5 + \frac{1}{T_{do}} \frac{x_d - x_d'}{x_{d_\Sigma}'} V_s \cos(x_3)$, $g_5(x) = \frac{1}{T_{do}}$, the state-space description of the power unit is written as

$$\dot{x}_1 = f_1(x) + g_1(x)x_2 \tag{2.96}$$

$$\dot{x}_2 = f_2(x) + g_2(x)u_1 \tag{2.97}$$

$$\dot{x}_3 = f_3(x) + g_3(x)x_4 \tag{2.98}$$

$$\dot{x}_4 = f_4(x) + g_4(x)x_5 \tag{2.99}$$

$$\dot{x}_5 = f_5(x) + g_5(x)u_2 \tag{2.100}$$

The system is differentially flat, with flat outputs $y_1 = x_1$ and $y_2 = x_3$. Besides, the system is in the triangular (backstepping integral) form. By considering a row-per-row decomposition of the system's dynamics, each row can be viewed as an independent sub0system. For the i-th row one has

$$\dot{x}_i = f_i(x) + g_i(x)x_{i+1} \tag{2.101}$$

where x_{i+1} can be considered to be a virtual control input v_i This subsystem is differentially flat with flat output x_i and a stabilizing feedback control about it is

$$v_i = \frac{1}{g_i(x)}[\dot{x}_{i,d} - f_i(x) - k_i(x_i - x_{i,d})] \tag{2.102}$$

The stabilizing feedback control for the system of Eq. (2.96) to Eq. (2.100) can be implemented in successive loops:

From Eq. (2.96), the setpoint of x_2 that achieves convergence of x_1 to the associated desirable output $x_{1,d}$ is

$$x_{2,d} = \frac{1}{g_1(x)}[\dot{x}_{1,d} - f_1(x) - k_1(x_1 - x_{1,d})] \tag{2.103}$$

From Eq. (2.97), the control input u_1 that achieves convergence of x_2 to the associated desirable output $x_{2,d}$ is given by

$$u_1 = \frac{1}{g_2(x)}[\dot{x}_{2,d} - f_2(x) - k_2(x_2 - x_{2,d})] \tag{2.104}$$

From Eq. (2.98), the setpoint of x_4 that achieves convergence of x_3 to the associated setpoint $x_{3,d}$ is

$$x_{4,d} = \frac{1}{g_3(x)}[\dot{x}_{3,d} - f_3(x) - k_3(x_3 - x_{3,d})] \tag{2.105}$$

From Eq. (2.99), the setpoint of x_5 that achieves convergence of x_4 to the associated setpoint $x_{4,d}$ is

$$x_{5,d} = \frac{1}{g_4(x)}[\dot{x}_{4,d} - f_4(x) - k_4(x_4 - x_{4,d})] \tag{2.106}$$

From Eq. (2.100), the control input u_2 that achieves convergence of x_5 to the associated desirable output $x_{5,d}$ is

$$u_2 = \frac{1}{g_5(x)}[\dot{x}_{5,d} - f_5(x) - k_5(x_5 - x_{5,d})] \tag{2.107}$$

By substituting Eq. (2.104) into Eq. (2.97) and after performing intermediate operations, one obtains

$$(\dot{x}_2 - \dot{x}_{2,d}) + k_2(x_2 - x_{2,d})] = 0 \Rightarrow \dot{e}_2 + k_2 e_2 = 0 \tag{2.108}$$

thus for $k_2 > 0$ one has

$$lim_{t \to \infty} e_2(t) = 0 \Rightarrow lim_{t \to \infty} x_2(t) = x_{2,d} \tag{2.109}$$

Using Eq. (2.109) and by substituting Eq. (2.104) into Eq. (2.96) one gets

$$(\dot{x}_1 - \dot{x}_{1,d}) + k_1(x_1 - x_{1,d}) = 0 \Rightarrow \dot{e}_1 + k_1 e_1 = 0 \tag{2.110}$$

thus for $k_1 > 0$ one has

$$lim_{t \to \infty} e_1(t) = 0 \Rightarrow lim_{t \to \infty} x_1(t) = x_{1,d} \tag{2.111}$$

Next, by substituting Eq. (2.107) into Eq. (2.100) one gets

$$(\dot{x}_5 - \dot{x}_{5,d}) + k_5(x_5 - x_{5,d})] = 0 \Rightarrow \dot{e}_5 + k_5 e_5 = 0 \tag{2.112}$$

thus for $k_5 > 0$ one has

$$lim_{t \to \infty} e_5(t) = 0 \Rightarrow lim_{t \to \infty} x_5(t) = x_{5,d} \tag{2.113}$$

Using Eq. (2.113) and by substituting Eq. (2.105) into Eq. (2.99) one gets

$$(\dot{x}_4 - \dot{x}_{4,d}) + k_4(x_4 - x_{4,d}) = 0 \Rightarrow \dot{e}_4 + k_4 e_4 = 0 \tag{2.114}$$

thus for $k_4 > 0$ one has

$$lim_{t \to \infty} e_4(t) = 0 \Rightarrow lim_{t \to \infty} x_4(t) = x_{4,d} \tag{2.115}$$

Using Eq. (2.115) and by substituting Eq. (2.105) into Eq. (2.99) one gets

$$(\dot{x}_3 - \dot{x}_{3,d}) + k_3(x_3 - x_{3,d})] = 0 \Rightarrow \dot{e}_3 + k_3 e_3 = 0 \tag{2.116}$$

thus for $k_3 > 0$ one has

$$lim_{t \to \infty} e_3(t) = 0 \Rightarrow lim_{t \to \infty} x_3(t) = x_{3,d} \tag{2.117}$$

The global stability properties of flatness-based control implemented in successive loops can be also proven with the use of the following Lyapunov function

$$V = \tfrac{1}{2}\sum_{i=1}^{5}(x_i - x_{i,d})^2 \Rightarrow$$
$$\dot{V} = \sum_{i=1}^{5}(x_i - x_{i,d})(\dot{x}_i - \dot{x}_{i,d}) \Rightarrow \qquad (2.118)$$
$$\dot{V} = \sum_{i=1}^{5} e_i \dot{e}_i$$

where for $i = 1, \cdots, 5$ it holds

$$\dot{e}_i + k_i e_i = 0 \Rightarrow \dot{e}_i = -k_i e_i \qquad (2.119)$$

By substituting Eq. (2.119) into Eq. (2.118), one obtains

$$\dot{V} = \sum_{i=1}^{5} e_i(-k_i e_i) \Rightarrow \dot{V} = -\sum_{i=1}^{5} k_i e_i^2 < 0 \qquad (2.120)$$

2.1.7 SLIDING-MODE CONTROL FOR MARINE-TURBINE AND SG POWER UNITS

The input-output linearized model of the marine turbine and synchronous generator power unit has been written in the form

$$\ddot{x}_1 = f_1(x) + g_{11}(x)u_1 + g_{21}(x)u_2 + \tilde{d}_1$$
$$x_3^{(3)} = f_2(x) + g_{12}(x)u_1 + g_{22}(x)u_2 + \tilde{d}_2 \qquad (2.121)$$

The following uncertainty regions are considered

$$|f_1 - \hat{f}_1| \leq \Delta F_1 \quad |f_2 - \hat{f}_2| \leq \Delta F_2 \qquad (2.122)$$

The following sliding surfaces are defined

$$s_1 = \dot{e}_1 + \lambda_{11} e_1 \quad s_2 = \ddot{e}_2 + \lambda_{21}\dot{e}_2 + \lambda_{22} e_2 \qquad (2.123)$$

where $e_1 = x_1 - x_{1,d}$ and $e_2 = x_3 - x_{3,d}$ are the tracking errors for state variables x_1 and x_3. The stabilizing feedback control is

$$u_1 = u_{1,eq} + u_{1,sw} \quad u_2 = u_{2,eq} + u_{2,sw} \qquad (2.124)$$

where $u_{i,eq}$, $i = 1, 2$ is the equivalent control term and $u_{i,sw}$, $i = 1, 2$ is the switching control term. Actually, the aggregate control for the marine-turbine and synchronous generator power unit is given by

$$\begin{pmatrix} u_1 \\ u_2 \end{pmatrix} = \begin{pmatrix} g_{11} & g_{12} \\ g_{21} & g_{22} \end{pmatrix}^{-1} \begin{pmatrix} \ddot{x}_{1,d} - \hat{f}_1 - \lambda_{11}\dot{e}_1 - k_1 sign(s_1) \\ x_{3,}^{(3)} - \hat{f}_2 - \lambda_{12}\ddot{e}_2 - \lambda_{22}\dot{e}_2 - k_2 sign(s_2) \end{pmatrix} \qquad (2.125)$$

Consequently, the equivalent control law and the switching control law for the power unit are

$$u_{eq} = \begin{pmatrix} g_{11} & g_{12} \\ g_{21} & g_{22} \end{pmatrix}^{-1} \begin{pmatrix} \ddot{x}_{1,d} - \hat{f}_1 - \lambda_{11}\dot{e}_1 \\ x_{3,}^{(3)} - \hat{f}_2 - \lambda_{12}\ddot{e}_2 - \lambda_{22}\dot{e}_2 \end{pmatrix}$$

$$u_{sw} = \begin{pmatrix} g_{11} & g_{12} \\ g_{21} & g_{22} \end{pmatrix}^{-1} \begin{pmatrix} -k_1 sign(s_1) \\ -k_2 sign(s_2) \end{pmatrix}$$

(2.126)

The equivalent control law is obtained from the following two conditions:

$$\dot{s}_1 = 0 \Rightarrow \ddot{e}_1 + \lambda_{11}\dot{e}_1 = 0 \Rightarrow$$
$$\ddot{x}_1 - \ddot{x}_{1,d} + \lambda_{11}(\dot{x}_1 - \dot{x}_{1,d}) = 0 \Rightarrow$$
$$f_1(x) + g_{11}(x)u_1 + g_{21}(x)u_2 - \ddot{x}_{1,d} + \lambda_{11}(\dot{x}_1 - \dot{x}_{1,d}) = 0 \Rightarrow$$
$$g_{11}(x)u_1 + g_{21}(x)u_2 = \ddot{x}_{1,d} - f_1(x) - \lambda_{11}(\dot{x}_1 - \dot{x}_{1,d})$$

(2.127)

and also

$$\dot{s}_2 = 0 \Rightarrow e_2^{(3)} + \lambda_{12}\ddot{e}_2 + \lambda_{22}\dot{e}_2 = 0 \Rightarrow$$
$$x_3^{(3)} - x_{3,d}^{(3)} + \lambda_{12}(\ddot{x}_3 - \ddot{x}_{3,d}) + \lambda_{22}(\dot{x}_3 - \dot{x}_{3,d}) = 0 \Rightarrow$$
$$f_2(x) + g_{12}(x)u_1 + g_{22}(x)u_2 - x_{3,d}^{(3)} + \lambda_{12}(\ddot{x}_3 - \ddot{x}_{3,d}) + \lambda_{22}(\dot{x}_3 - \dot{x}_{3,d}) = 0 \Rightarrow$$
$$g_{12}(x)u_1 + g_{22}(x)u_2 = x_{3,d}^{(3)} - f_2(x) - \lambda_{12}(\ddot{x}_3 - \ddot{x}_{3,d}) - \lambda_{22}(\dot{x}_3 - \dot{x}_{3,d})$$

(2.128)

From Eq. (2.127) and Eq. (2.128) one has

$$\begin{pmatrix} g_{11}(x) & g_{12}(x) \\ g_{21}(x) & g_{22}(x) \end{pmatrix} \begin{pmatrix} u_{1,eq} \\ u_{2,eq} \end{pmatrix} = \begin{pmatrix} \ddot{x}_{1,d} - f_1(x) - \lambda_{11}(\dot{x}_1 - \dot{x}_{1,d}) \\ x_{3,d}^{(3)} - f_2(x) - \lambda_{12}(\ddot{x}_3 - \ddot{x}_{3,d}) - \lambda_{22}(\dot{x}_3 - \dot{x}_{3,d}) \end{pmatrix}$$

(2.129)

and since $f_1(x)$, $f_2(x)$ are known within uncertainty ranges they are substituted by their estimated values $\hat{f}_1(x)$, $\hat{f}_2(x)$, the previous relation becomes

$$\begin{pmatrix} u_{1,eq} \\ u_{2,eq} \end{pmatrix} = \begin{pmatrix} g_{11}(x) & g_{12}(x) \\ g_{21}(x) & g_{22}(x) \end{pmatrix}^{-1} \begin{pmatrix} \ddot{x}_{1,d} - \hat{f}_1(x) - \lambda_{11}(\dot{x}_1 - \dot{x}_{1,d}) \\ x_{3,d}^{(3)} - \hat{f}_2(x) - \lambda_{12}(\ddot{x}_3 - \ddot{x}_{3,d}) - \lambda_{22}(\dot{x}_3 - \dot{x}_{3,d}) \end{pmatrix}$$

(2.130)

To compensate for model uncertainties, the switching control terms are included in the control inputs. Thus, the aggregate control inputs become

$$\begin{pmatrix} u_1 \\ u_2 \end{pmatrix} = \begin{pmatrix} g_{11}(x) & g_{12}(x) \\ g_{21}(x) & g_{22}(x) \end{pmatrix}^{-1} \cdot$$

$$\cdot \begin{pmatrix} \ddot{x}_{1,d} - \hat{f}_1(x) - \lambda_{11}(\dot{x}_1 - \dot{x}_{1,d}) - k_1 sign(s_1) \\ x_{3,d}^{(3)} - \hat{f}_2(x) - \lambda_{12}(\ddot{x}_3 - \ddot{x}_{3,d}) - \lambda_{22}(\dot{x}_3 - \dot{x}_{3,d}) - k_2 sign(s_2) \end{pmatrix}$$

(2.131)

The stability conditions of the sliding-mode control scheme are as follows:

$$\frac{1}{2}\frac{d}{dt}s_i^2 \leq -\eta_i|s_i| \quad i = 1,2 \Rightarrow \tag{2.132}$$

$$s_1\dot{s}_1 \leq -\eta_1|s_1| \Rightarrow s_1[\ddot{e}_1 + \lambda_{11}\dot{e}_1] \leq -\eta_1|s_1|$$
$$s_2\dot{s}_2 \leq -\eta_2|s_1| \Rightarrow s_2[e_3^{(3)} + \lambda_{12}\ddot{e}_2 + \lambda_{22}\dot{e}_2] \leq -\eta_2|s_2| \tag{2.133}$$

Equivalently, one obtains

$$s_1[f_1(x) + g_{11}(x)u_1 + g_{12}(x)u_2 - \ddot{x}_{1,d} + \lambda_{11}(\dot{x}_1 - \dot{x}_{1,d})] \leq -\eta_1|s_1|$$
$$s_2[f_2(x) + g_{12}(x)u_2 + g_{22}(x)u_2 - x_{3,d}^{(3)} + \lambda_{12}(\ddot{x}_3 - \ddot{x}_{3,d}) + \lambda_{22}(\dot{x}_3 - \dot{x}_{3,d})] \leq -\eta_2|s_2| \tag{2.134}$$

Consequently, one obtains

$$\begin{pmatrix} s_1 & 0 \\ 0 & s_2 \end{pmatrix} \left\{ \begin{pmatrix} f_1(x) \\ f_2(x) \end{pmatrix} + \begin{pmatrix} g_{11}(x) & g_{12}(x) \\ g_{21}(x) & g_{22}(x) \end{pmatrix} \begin{pmatrix} u_1 \\ u_2 \end{pmatrix} + \right.$$
$$\left. + \begin{pmatrix} -\ddot{x}_{1,d} + \lambda_{11}(\dot{x}_1 - \dot{x}_{1,d}) \\ -x_{3,d}^{(3)} + \lambda_{12}(\ddot{x}_3 - \ddot{x}_{3,d}) + \lambda_{22}(\dot{x}_2 - \dot{x}_{2,d}) \end{pmatrix} \right\} \leq - \begin{pmatrix} \eta_1|s_1| \\ \eta_2|s_2| \end{pmatrix} \tag{2.135}$$

Equivalently, one has

$$\begin{pmatrix} s_1 & 0 \\ 0 & s_2 \end{pmatrix} \left\{ \begin{pmatrix} f_1(x) \\ f_2(x) \end{pmatrix} + \begin{pmatrix} \ddot{x}_{1,d} - \hat{f}_1(x) - \lambda_{11}(\dot{x}_1 - \dot{x}_{1,d}) - k_1 sign(x_1) \\ x_{3,d}^{(3)} - \hat{f}_2(x) - \lambda_{12}(\ddot{x}_3 - \ddot{x}_{3,d}) - \lambda_{12}(\dot{x}_3 - \dot{x}_{3,d}) - k_2 sign(s_2) \end{pmatrix} + \right.$$
$$\left. + \begin{pmatrix} -\ddot{x}_{1,d} + \lambda_{11}(\dot{x}_1 - \dot{x}_{1,d}) \\ -x_{3,d}^{(3)} + \lambda_{12}(\ddot{x}_3 - \ddot{x}_{3,d}) + \lambda_{22}(\dot{x}_2 - \dot{x}_{2,d}) \end{pmatrix} \right\} \leq - \begin{pmatrix} \eta_1|s_1| \\ \eta_2|s_2| \end{pmatrix} \tag{2.136}$$

After intermediate operations, one obtains

$$\begin{pmatrix} s_1 & 0 \\ 0 & s_2 \end{pmatrix} \left\{ \begin{pmatrix} f_1(x) - \hat{f}_1(x) \\ f_2(x) - \hat{f}_2(x) \end{pmatrix} - \begin{pmatrix} k_1 sign(x_1) \\ k_2 sign(s_2) \end{pmatrix} \right\} \leq - \begin{pmatrix} \eta_1|s_1| \\ \eta_2|s_2| \end{pmatrix} \tag{2.137}$$

or equivalently

$$\begin{pmatrix} s_1 & 0 \\ 0 & s_2 \end{pmatrix} \left\{ \begin{pmatrix} \Delta f_1(x) - k_1 sign(x_1) \\ \Delta f_2(x) - k_2 sign(s_2) \end{pmatrix} \right\} \leq - \begin{pmatrix} \eta_1|s_1| \\ \eta_2|s_2| \end{pmatrix} \tag{2.138}$$

Next, one obtains

$$\Delta F_1 s_1 - k_1 sign(s_1)s_1 \leq -\eta_1|s_1|$$
$$\Delta F_2 s_2 - k_2 sign(s_2)s_2 \leq -\eta_2|s_2| \tag{2.139}$$

and after additional operations one has

$$\begin{aligned}\Delta F_1 s_1 - k_1|s_1| &\leq -\eta_1|s_1| \\ \Delta F_2 s_2 - k_2|s_2| &\leq -\eta_2|s_2|\end{aligned} \qquad (2.140)$$

By setting $k_1 = \eta_1 + \Delta F_1$ and $k_2 = \eta_2 + \Delta F_2$ one gets

$$\begin{aligned}\Delta F_1 s_1 - \Delta F_1|s_1| - \eta_1|s_1| &\leq -\eta_1|s_1| \\ \Delta F_2 s_2 - \Delta F_2|s_2| - \eta_2|s_2| &\leq -\eta_2|s_2|\end{aligned} \qquad (2.141)$$

which finally gives

$$\begin{aligned}\Delta F_1 s_1 &\leq \Delta F_1|s_1| \\ \Delta F_2 s_2 &\leq \Delta F_2|s_2|\end{aligned} \qquad (2.142)$$

which is a condition that always holds. Consequently, through the selection of the switching control gains k_1, k_2 given above, one ensures that the global asymptotic stability condition for the sliding-mode control scheme always holds.

2.1.8   MULTI-MODEL FUZZY CONTROL FOR MARINE-TURBINE AND SG POWER UNITS

The Jacobian matrices of the linearized state-space model of the marine-turbine and synchronous generator power unit are used as given in Eq. (2.10) and Eq. (2.11). Using also that $x_{5,d}$ is the desirable value for the fifth element of the state vector of the power unit, linearization is now performed around the following fixed operating points: $(x_3^*, x_5^*) = (0, x_5^d)$, $(x_3^*, x_5^*) = (\frac{\pi}{4}, x_5^d)$, $(x_3^*, x_5^*) = (\frac{\pi}{2}, x_5^d)$, $(x_3^*, x_5^*) = (\frac{3\pi}{4}, x_5^d)$, $(x_3^*, x_5^*) = (\pi, x_5^d)$, $(x_3^*, x_5^*) = (\frac{5\pi}{4}, x_5^d)$, $(x_3^*, x_5^*) = (\frac{3\pi}{2}, x_5^d)$, and $(x_3^*, x_5^*) = (\frac{7\pi}{4}, x_5^d)$.

Matrix $B = \nabla_u[f(x) + g(x)u]$ computed at the above noted linearization points remains as in Eq. (2.8). Matrix $A = \nabla_x[f(x) + g(x)u]$ has the generic form

$$A = \begin{pmatrix} 0 & 1 & 0 & 0 & 0 \\ -\frac{K_1}{J_t} & -\frac{D_1}{J_t} & \frac{K_1}{J_t} & 0 & 0 \\ 0 & 0 & 0 & 1 & 0 \\ \frac{K_1}{J_g}\frac{n_g}{n_t} & 0 & -\frac{K_1}{J_g}\frac{n_g}{n_t} & \frac{\omega_0 V_s x_5^* \cos(x_3^*)}{x_{d\Sigma}'} & -\frac{D_g}{J_g} & -\frac{\omega_0 V_s \sin(x_3^*)}{x_{d\Sigma}'} \\ 0 & 0 & -\frac{1}{T_{do}'}\frac{x_d - x_d'}{x_{d\Sigma}'}V_s \sin(x_3^*) & 0 & -\frac{1}{T_{[d]}'} \end{pmatrix} \qquad (2.143)$$

At linearization point $(x_3^*, x_5^*) = (0, x_5^d)$ the elements of matrix A become: $a_{43} = -\frac{K_1}{J_g}\frac{n_g}{n_t} - \frac{\omega_0 V_s x_5^*}{x_{d\Sigma}'}$, $a_{45} = 0$, $a_{53} = 0$.

At linearization point $(x_3^*, x_5^*) = (\frac{\pi}{4}, x_5^d)$ the elements of matrix A become: $a_{43} = -\frac{K_1}{J_g}\frac{n_g}{n_t} - \frac{\omega_0 V_s x_5^*\frac{\sqrt{2}}{2}}{x_{d\Sigma}'}$, $a_{45} = -\frac{\omega_0 V_s\frac{\sqrt{2}}{2}}{x_{d\Sigma}'}$, $a_{53} = -\frac{1}{T_{do}'}\frac{x_d - x_d'}{x_{d\Sigma}'}V_s\frac{\sqrt{2}}{2}$.

At linearization point $(x_3^*, x_5^*) = (\frac{\pi}{2}, x_5^d)$ the elements of matrix A become: $a_{43} = -\frac{K_1}{J_g}\frac{n_g}{n_t}$, $a_{45} = -\frac{\omega_0 V_s}{x_{d\Sigma}'}$, $a_{53} = -\frac{1}{T_{do}'}\frac{x_d - x_d'}{x_{d\Sigma}'}V_s$.

At linearization point $(x_3^*, x_5^*) = (\frac{3\pi}{4}, x_5^d)$ the elements of matrix A become: $a_{43} = -\frac{K_1}{J_g}\frac{n_g}{n_t} + \frac{\omega_0 V_s x_5^* \frac{\sqrt{2}}{2}}{x_{d\Sigma}'}$, $a_{45} = -\frac{\omega_0 V_s \frac{\sqrt{2}}{2}}{x_{d\Sigma}'}$, $a_{53} = -\frac{1}{T_{do}'}\frac{x_d - x_d'}{x_{d\Sigma}'}V_s\frac{\sqrt{2}}{2}$.

At linearization point $(x_3^*, x_5^*) = (\pi, x_5^d)$ the elements of matrix A become: $a_{43} = -\frac{K_1}{J_g}\frac{n_g}{n_t} + \frac{\omega_0 V_s x_5^*}{x_{d\Sigma}'}$, $a_{45} = 0$, $a_{53} = 0$.

At linearization point $(x_3^*, x_5^*) = (\frac{5\pi}{4}, x_5^d)$ the elements of matrix A become: $a_{43} = -\frac{K_1}{J_g}\frac{n_g}{n_t} + \frac{\omega_0 V_s x_5^* \frac{\sqrt{2}}{2}}{x_{d\Sigma}'}$, $a_{45} = \frac{\omega_0 V_s \frac{\sqrt{2}}{2}}{x_{d\Sigma}'}$, $a_{53} = \frac{1}{T_{do}'}\frac{x_d - x_d'}{x_{d\Sigma}'}V_s\frac{\sqrt{2}}{2}$.

At linearization point $(x_3^*, x_5^*) = (\frac{3\pi}{2}, x_5^d)$ the elements of matrix A become: $a_{43} = -\frac{K_1}{J_g}\frac{n_g}{n_t}$, $a_{45} = \frac{\omega_0 V_s}{x_{d\Sigma}'}$, $a_{53} = \frac{1}{T_{do}'}\frac{x_d - x_d'}{x_{d\Sigma}'}V_s$.

At linearization point $(x_3^*, x_5^*) = (\frac{7\pi}{4}, x_5^d)$. the elements of matrix A become: $a_{43} = -\frac{K_1}{J_g}\frac{n_g}{n_t} - \frac{\omega_0 V_s x_5^* \frac{\sqrt{2}}{2}}{x_{d\Sigma}'}$, $a_{45} = \frac{\omega_0 V_s \frac{\sqrt{2}}{2}}{x_{d\Sigma}'}$, $a_{53} = \frac{1}{T_{do}'}\frac{x_d - x_d'}{x_{d\Sigma}'}V_s\frac{\sqrt{2}}{2}$.

Taking into account matrices A_i $i = 1, \cdots, 8$ which are obtained from the linearization process, the dynamic model of the marine-turbine and synchronous generator power unit is described by the following eight fuzzy rules:

R_1: IF $x_3 \in A_1$ (where A_1 is a fuzzy set centered at $x_3 = 0$) THEN the model of the power unit is $\dot{x} = A_1 x + Bu$ AND the stabilizing feedback control is $u_1 = K_1 e$

R_2: IF $x_3 \in A_2$ (where A_2 is a fuzzy set centered at $x_3 = \frac{\pi}{4}$) THEN the model of the power unit is $\dot{x} = A_2 x + Bu$ AND the stabilizing feedback control is $u_2 = K_2 e$

R_3: IF $x_3 \in A_3$ (where A_3 is a fuzzy set centered at $x_3 = \frac{\pi}{2}$) THEN the model of the power unit is $\dot{x} = A_3 x + Bu$ AND the stabilizing feedback control is $u_3 = K_3 e$

R_4: IF $x_3 \in A_4$ (where A_4 is a fuzzy set centered at $x_3 = \frac{3\pi}{4}$) THEN the model of the power unit is $\dot{x} = A_4 x + Bu$ AND the stabilizing feedback control is $u_4 = K_4 e$

R_5: IF $x_3 \in A_5$ (where A_5 is a fuzzy set centered at $x_3 = \pi$) THEN the model of the power unit is $\dot{x} = A_5 x + Bu$ AND the stabilizing feedback control is $u_5 = K_5 e$

R_6: IF $x_3 \in A_6$ (where A_6 is a fuzzy set centered at $x_3 = \frac{5\pi}{4}$) THEN the model of the power unit is $\dot{x} = A_6 x + Bu$ AND the stabilizing feedback control is $u_6 = K_6 e$

R_7: IF $x_3 \in A_7$ (where A_7 is a fuzzy set centered at $x_3 = \frac{3\pi}{2}$) THEN the model of the power unit is $\dot{x} = A_7 x + Bu$ AND the stabilizing feedback control is $u_7 = K_7 e$

R_8: IF $x_3 \in A_8$ (where A_8 is a fuzzy set centered at $x_3 = \frac{7\pi}{4}$) THEN the model of the power unit is $\dot{x} = A_8 x + B u$ AND the stabilizing feedback control is $u_8 = K_8 e$

The stabilizing feedback gains K_j $j = 1, \cdots, 8$ of the local controllers $u_j = K_j e$ $j = 1, \cdots 8$ can be obtained from the solution of the associated H-infinity control problem. Such gains are given by $K = -\frac{1}{r_j} B^T P_j$ where the positive definite and symmetric matrix P_j $j = 1, \cdots, 8$ is obtained from the solution of an algebraic Riccati equation of the form of Eq. (2.26).

The aggregate control signal that is applied to the marine-turbine and synchronous generator power unit becomes

$$u(t) = \frac{\sum_{j=1}^{8} w_j K_j e(t)}{\sum_{j=1}^{8} w_j} \tag{2.144}$$

where w_j is the membership value of x_3 in fuzzy set A_j. By applying the above feedback control the output of the closed-loop system becomes

$$\delta x(t) = \frac{\sum_{i=1}^{8} \sum_{j=1}^{8} w_i w_j (A_i + B_i K_j) e(t)}{\sum_{i=1}^{8} \sum_{j=1}^{8} w_i w_j} \tag{2.145}$$

Stability conditions for this continuous-time Takagi-Sugeno fuzzy system under state feedback can be also formulated [195].

Condition 1: The equilibrium of the fuzzy system is globally asymptotically stable if there exists a common symmetric positive definite matrix P such that

$$(A_i + BK_j)^T P + P(A_i + BK_j) < 0 \; \forall \; i, j = 1, 2, \cdots, 8 \tag{2.146}$$

Finally, a less conservative condition about the stability of the continuous-time Takagi-Sugeno system is formulated as follows [195]:

Condition 2: The equilibrium state for the continuous-time fuzzy system is globally asymptotically stable if there exists a common positive definite matrix P such that

$$\begin{aligned} (A_i + BK_i)^T P + P(A_i + BK_i) < 0, \; i = 1, 2, \cdots, 8 \\ (G_{ij}^T P + PG_{ij}) < 0, \; i < j \leq 8 \end{aligned} \tag{2.147}$$

where

$$G_{ij} = \frac{(A_i + BK_j) + (A_j + BK_i)}{2}, \; i, j \leq 8 \tag{2.148}$$

2.1.9 SIMULATION TESTS

2.1.9.1 Results on nonlinear optimal control

The efficiency of the proposed nonlinear optimal (H-infinity) control scheme has been tested through simulation experiments. The functioning of a stand-alone marine power unit that comprises one marine turbine connected to one synchronous generator has been examined. As commonly established for the dynamic model of the Single-Machine Infinite Bus (SMIB) power system, state variables of the model are rated (normalized) and are expressed in the per unit (p.u.) system. The obtained results are given in Fig. 2.2 to Fig. 2.6 and demonstrate that under nonlinear optimal control fast and accurate tracking of the reference setpoints was achieved by the state variables of the marine power unit. Moreover, the variations of the control inputs were moderate. The real values of the state variables are depicted in blue, the estimated variables (provided by an H-infinity Kalman Filter) are printed in green, while the associated reference setpoints are plotted in red.

2.1.9.2   Results on Lie algebra-based control

The obtained results about Lie-algebra-based control in the model of the marine power unit are given in Fig. 2.7 to Fig. 2.11, and confirm that this method performs fine. Unlike global linearization-based control approaches, such as Lie algebra-based control, the optimal control approach does not rely on complicated transformations (diffeomorphisms) of the system's state variables. Besides, the computed control inputs are applied directly on the initial nonlinear model of the marine-turbine and synchronous generator power system and not on its linearized equivalent description. In case of nonlinear optimal control, the inverse transformations which are met in global linearization-based control are avoided and consequently one does not come against the related singularity problems.

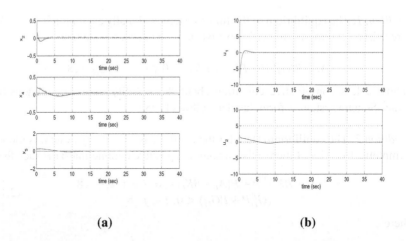

(a)   (b)

Figure 2.2 Test 1 of nonlinear optimal control for the marine-turbine power unit: (a) Convergence of state variables x_2, x_4 and x_5 (blue lines) to the reference setpoints (red lines) and state estimates provided by the Kalman Filter (green lines) and (b) control inputs u_1, u_2 applied to the marine-power generation system

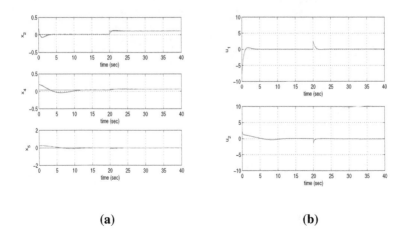

(a) (b)

Figure 2.3 Test 2 of nonlinear optimal control for the marine-turbine power unit: (a) Convergence of state variables x_2, x_4 and x_5 (blue lines) to the reference setpoints (red lines) and state estimates provided by the Kalman Filter (green lines) and (b) control inputs u_1, u_2 applied to the marine-power generation system

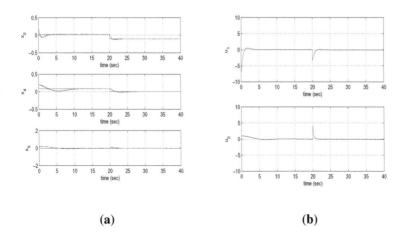

(a) (b)

Figure 2.4 Test 3 of nonlinear optimal control for the marine-turbine power unit: (a) Convergence of state variables x_2, x_4 and x_5 (blue lines) to the reference setpoints (red lines) and state estimates provided by the Kalman Filter (green lines) and (b) control inputs u_1, u_2 applied to the marine-power generation system

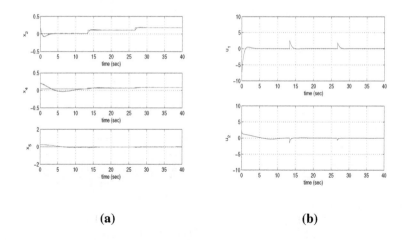

(a)   (b)

Figure 2.5   Test 4 of nonlinear optimal control for the marine-turbine power unit: (a) Convergence of state variables x_2, x_4 and x_5 (blue lines) to the reference setpoints (red lines) and state estimates provided by the Kalman Filter (green lines) and (b) control inputs u_1, u_2 applied to the marine-power generation system

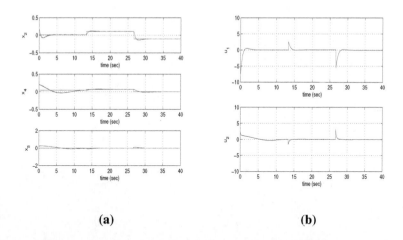

(a)   (b)

Figure 2.6   Test 5 of nonlinear optimal control for the marine-turbine power unit: (a) Convergence of state variables x_2, x_4 and x_5 (blue lines) to the reference setpoints (red lines) and state estimates provided by the Kalman Filter (green lines) and (b) control inputs u_1, u_2 applied to the marine-power generation system

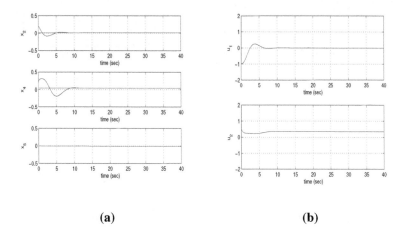

(a) (b)

Figure 2.7 Test 1 of Lie algebra-based control with transformation into canonical forms for the marine-turbine power unit: (a) Convergence of state variables x_2, x_4 and x_5 (blue lines) to the reference setpoints (red lines) and (b) control inputs u_1, u_2 applied to the marine-power generation system

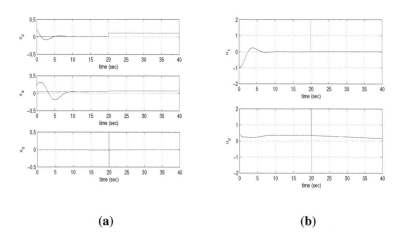

(a) (b)

Figure 2.8 Test 2 of Lie algebra-based control with transformation into canonical forms for the marine-turbine power unit: (a) Convergence of state variables x_2, x_4 and x_5 (blue lines) to the reference setpoints (red lines) and (b) control inputs u_1, u_2 applied to the marine-power generation system

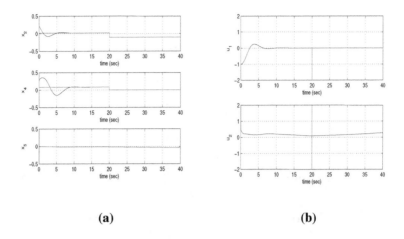

(a)   **(b)**

Figure 2.9 Test 3 of Lie algebra-based control with transformation into canonical forms for the marine-turbine power unit: (a) Convergence of state variables x_2, x_4 and x_5 (blue lines) to the reference setpoints (red lines) and (b) control inputs u_1, u_2 applied to the marine-power generation system

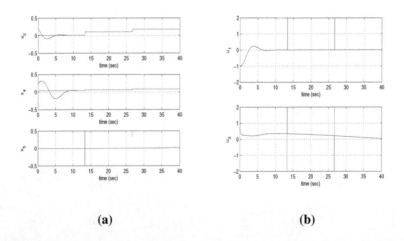

(a)   **(b)**

Figure 2.10 Test 4 of Lie algebra-based control with transformation into canonical forms for the marine-turbine power unit: (a) Convergence of state variables x_2, x_4 and x_5 (blue lines) to the reference setpoints (red lines) and (b) control inputs u_1, u_2 applied to the marine-power generation system

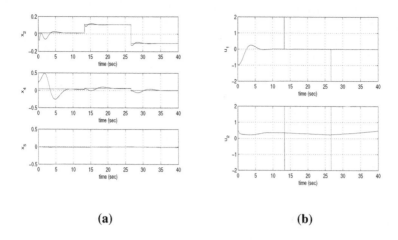

(a)   (b)

Figure 2.11 Test 5 of Lie algebra-based control with transformation into canonical forms for the marine-turbine power unit: (a) Convergence of state variables x_2, x_4 and x_5 (blue lines) to the reference setpoints (red lines) and (b) control inputs u_1, u_2 applied to the marine-power generation system

2.1.9.3 Results on flatness-based control with transformation into canonical forms

The obtained results about flatness-based control with transformation of the model of the marine power unit in canonical (Brunovsky) form are given in Fig. 2.12 to Fig. 2.39, and confirm that this method performs fine. Unlike global linearization-based control approaches, such as flatness-based control with transformation into canonical forms, the previously given optimal control approach does not rely on complicated transformations (diffeomorphisms) of the system's state variables. Besides, in the optimal control approach the computed control inputs are applied directly on the initial nonlinear model of the marine-turbine and synchronous generator power system and not on its linearized equivalent. Moreover, in nonlinear optimal control the inverse transformations which are met in global linearization-based control are avoided and consequently one does not come against the related singularity problems.

2.1.9.4 Results on sliding-mode control

The obtained results about sliding-mode control in the model of the marine power unit are given in Fig. 2.22 to Fig. 2.26, and confirm that this method performs fine. Unlike sliding mode control and variable structure control schemes, the previously given optimal control method does not require the state-space description of the system to be found in a specific form. About sliding-mode control it is known that when the controlled system is not found in the input-output linearized form the definition of the sliding surface can be an intuitive procedure.

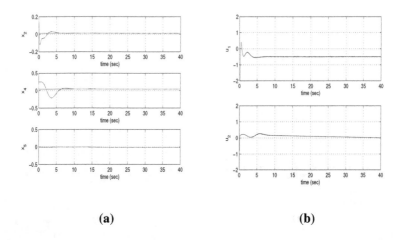

(a)   **(b)**

Figure 2.12   Test 1 of flatness-based control with transformation into canonical forms for the marine-turbine power unit: (a) Convergence of state variables x_2, x_4 and x_5 (blue lines) to the reference setpoints (red lines) and (b) control inputs u_1, u_2 applied to the marine-power generation system

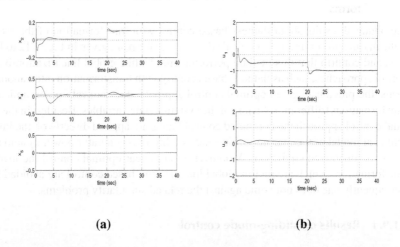

(a)   **(b)**

Figure 2.13   Test 2 of flatness-based control with transformation into canonical forms for the marine-turbine power unit: (a) Convergence of state variables x_2, x_4 and x_5 (blue lines) to the reference setpoints (red lines) and (b) control inputs u_1, u_2 applied to the marine-power generation system

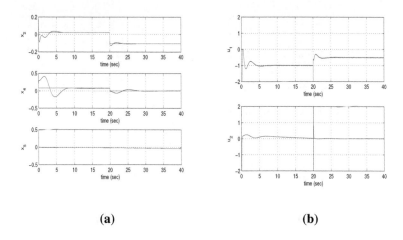

(a) **(b)**

Figure 2.14 Test 3 of flatness-based control with transformation into canonical forms for the marine-turbine power unit: (a) Convergence of state variables x_2, x_4 and x_5 (blue lines) to the reference setpoints (red lines) and (b) control inputs u_1, u_2 applied to the marine-power generation system

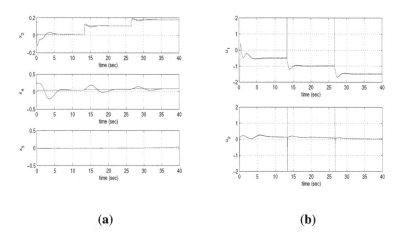

(a) **(b)**

Figure 2.15 Test 4 of flatness-based control with transformation into canonical forms for the marine-turbine power unit: (a) Convergence of state variables x_2, x_4 and x_5 (blue lines) to the reference setpoints (red lines) and (b) control inputs u_1, u_2 applied to the marine-power generation system

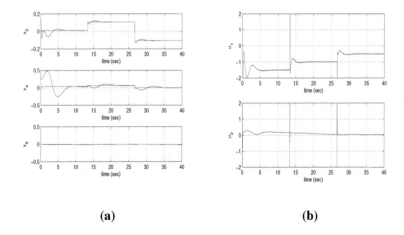

(a)   **(b)**

Figure 2.16   Test 5 of flatness-based control with transformation into canonical forms for the marine-turbine power unit: (a) Convergence of state variables x_2, x_4 and x_5 (blue lines) to the reference setpoints (red lines) and (b) control inputs u_1, u_2 applied to the marine-power generation system

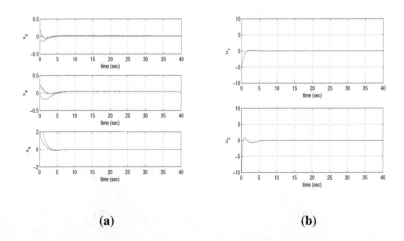

(a)   **(b)**

Figure 2.17   Test 1 of flatness-based control in successive loops for the marine-turbine power unit: (a) Convergence of state variables x_2, x_4 and x_5 (blue lines) to the reference setpoints (red lines) and (b) control inputs u_1, u_2 applied to the marine-power generation system

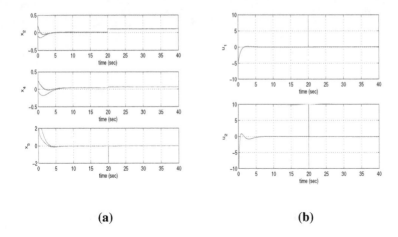

(a) (b)

Figure 2.18 Test 2 of flatness-based control in successive loops for the marine-turbine power unit: (a) Convergence of state variables x_2, x_4 and x_5 (blue lines) to the reference setpoints (red lines) and (b) control inputs u_1, u_2 applied to the marine-power generation system

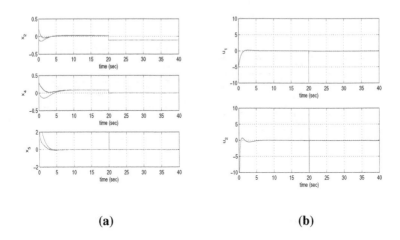

(a) (b)

Figure 2.19 Test 3 of flatness-based control in successive loops for the marine-turbine power unit: (a) Convergence of state variables x_2, x_4 and x_5 (blue lines) to the reference setpoints (red lines) and (b) control inputs u_1, u_2 applied to the marine-power generation system

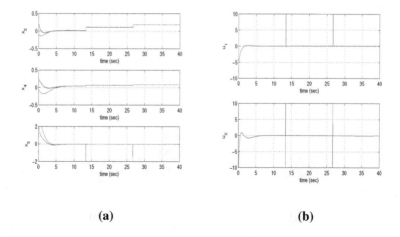

(a)   (b)

Figure 2.20 Test 4 of flatness-based control in successive loops for the marine-turbine power unit: (a) Convergence of state variables x_2, x_4 and x_5 (blue lines) to the reference setpoints (red lines) and (b) control inputs u_1, u_2 applied to the marine-power generation system

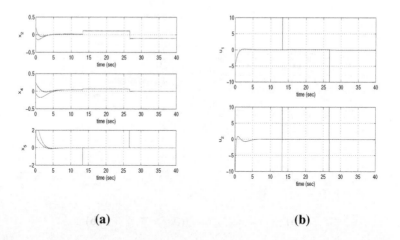

(a)   (b)

Figure 2.21 Test 5 of flatness-based control in successive loops for the marine-turbine power unit: (a) Convergence of state variables x_2, x_4 and x_5 (blue lines) to the reference setpoints (red lines) and (b) control inputs u_1, u_2 applied to the marine-power generation system

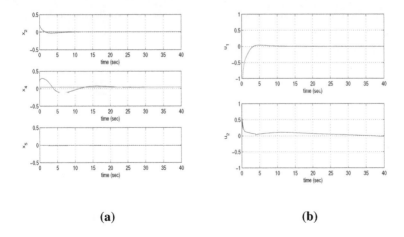

Figure 2.22 Test 1 of sliding-mode control (with saturation function) for the marine-turbine power unit: (a) Convergence of state variables x_2, x_4 and x_5 (blue lines) to the reference set-points (red lines) and (b) control inputs u_1, u_2 applied to the marine-power generation system

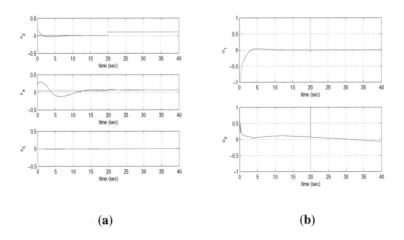

Figure 2.23 Test 2 of sliding-mode control (with saturation function) for the marine-turbine power unit: (a) Convergence of state variables x_2, x_4 and x_5 (blue lines) to the reference set-points (red lines) and (b) control inputs u_1, u_2 applied to the marine-power generation system

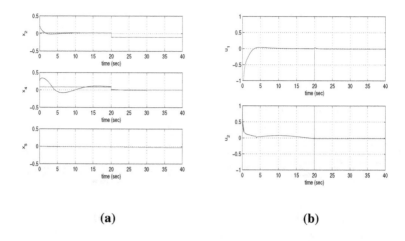

(a)   (b)

Figure 2.24   Test 3 of sliding-mode control (with saturation function)for the marine-turbine power unit: (a) Convergence of state variables x_2, x_4 and x_5 (blue lines) to the reference set-points (red lines) and (b) control inputs u_1, u_2 applied to the marine-power generation system

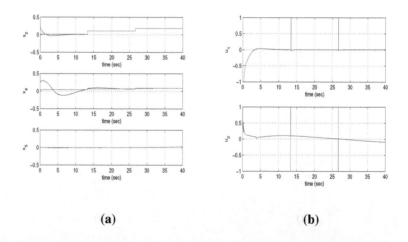

(a)   (b)

Figure 2.25   Test 4 of sliding-mode control (with saturation function) for the marine-turbine power unit: (a) Convergence of state variables x_2, x_4 and x_5 (blue lines) to the reference set-points (red lines) and (b) control inputs u_1, u_2 applied to the marine-power generation system

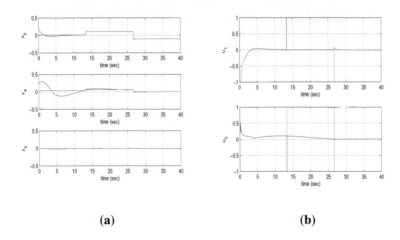

(a)   (b)

Figure 2.26 Test 5 of sliding-mode control (with saturation function) for the marine-turbine power unit: (a) Convergence of state variables x_2, x_4 and x_5 (blue lines) to the reference set-points (red lines) and (b) control inputs u_1, u_2 applied to the marine-power generation system

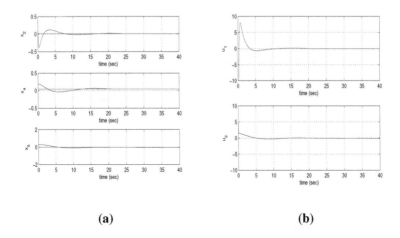

(a)   (b)

Figure 2.27 Test 1 of fuzzy multiple model-based control in successive loops for the marine-turbine power unit: (a) Convergence of state variables x_2, x_4 and x_5 (blue lines) to the reference setpoints (red lines) and (b) control inputs u_1, u_2 applied to the marine-power generation system

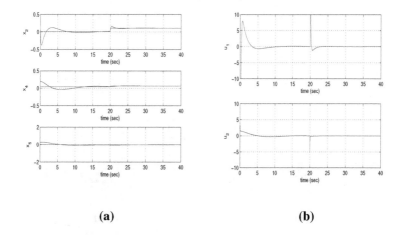

(a)   **(b)**

Figure 2.28   Test 2 of fuzzy multiple model-based control in successive loops for the marine-turbine power unit: (a) Convergence of state variables x_2, x_4 and x_5 (blue lines) to the reference setpoints (red lines) and (b) control inputs u_1, u_2 applied to the marine-power generation system

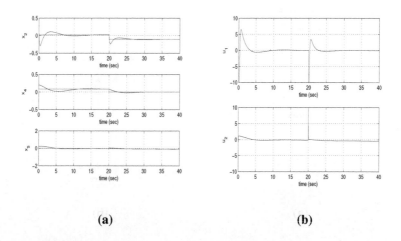

(a)   **(b)**

Figure 2.29   Test 3 of fuzzy multiple models-based control in successive loops for the marine-turbine power unit: (a) Convergence of state variables x_2, x_4 and x_5 (blue lines) to the reference setpoints (red lines) and (b) control inputs u_1, u_2 applied to the marine-power generation system

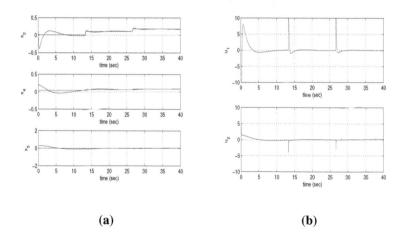

(a)   **(b)**

Figure 2.30 Test 4 of fuzzy multiple model-based control in successive loops for the marine-turbine power unit: (a) Convergence of state variables x_2, x_4 and x_5 (blue lines) to the reference setpoints (red lines) and (b) control inputs u_1, u_2 applied to the marine-power generation system

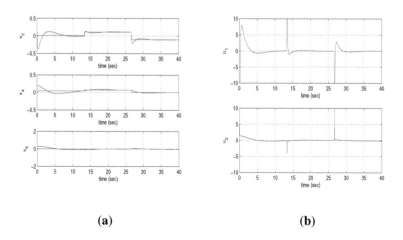

(a)   **(b)**

Figure 2.31 Test 5 of fuzzy multiple model-based control in successive loops for the marine-turbine power unit: (a) Convergence of state variables x_2, x_4 and x_5 (blue lines) to the reference setpoints (red lines) and (b) control inputs u_1, u_2 applied to the marine-power generation system

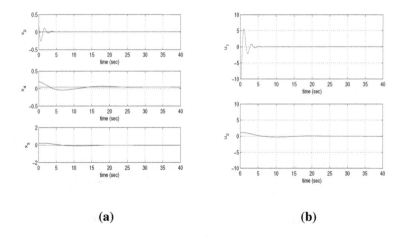

(a)   (b)

Figure 2.32   Test 1 of PID control (in two independent loops) for the marine-turbine power unit: (a) Convergence of state variables x_2, x_4 and x_5 (blue lines) to the reference setpoints (red lines) and (b) control inputs u_1, u_2 applied to the marine-power generation system

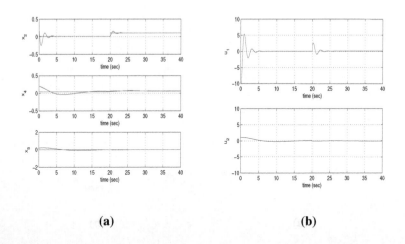

(a)   (b)

Figure 2.33   Test 2 of PID control (in two independent loops) for the marine-turbine power unit: (a) Convergence of state variables x_2, x_4 and x_5 (blue lines) to the reference setpoints (red lines) and (b) control inputs u_1, u_2 applied to the marine-power generation system

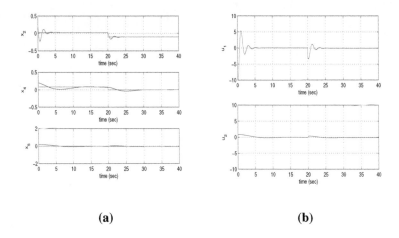

(a)   (b)

Figure 2.34 Test 3 of PID control (in two independent loops) for the marine-turbine power unit: (a) Convergence of state variables x_2, x_4 and x_5 (blue lines) to the reference setpoints (red lines) and (b) control inputs u_1, u_2 applied to the marine-power generation system

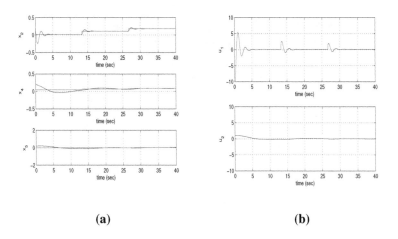

(a)   (b)

Figure 2.35 Test 4 of PID control (in two independent loops) for the marine-turbine power unit: (a) Convergence of state variables x_2, x_4 and x_5 (blue lines) to the reference setpoints (red lines) and (b) control inputs u_1, u_2 applied to the marine-power generation system

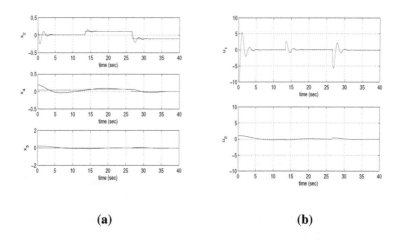

(a)   (b)

Figure 2.36 Test 5 of PID control (in two independent loops) for the marine-turbine power unit: (a) Convergence of state variables x_2, x_4 and x_5 (blue lines) to the reference setpoints (red lines) and (b) control inputs u_1, u_2 applied to the marine-power generation system

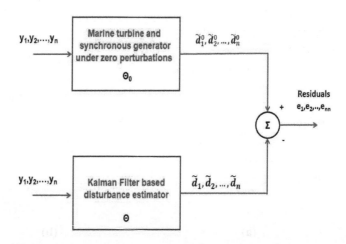

Figure 2.37 Residuals'generation for the marine-turbine and synchronous generators power unit, with a Kalman Filter-based disturbance estimator and through the processing of the estimated disturbance inputs

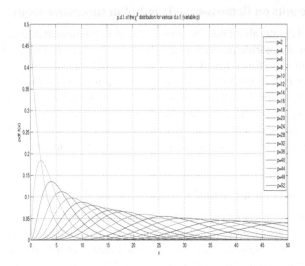

Figure 2.38 Probability density function of the χ^2 distribution for several values of the degrees of freedom (given by variable p)

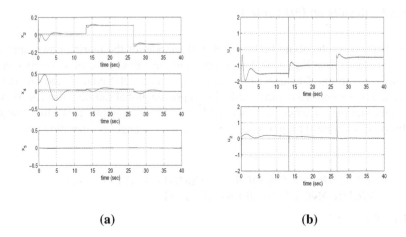

(a)   **(b)**

Figure 2.39 Indicative test of flatness-based control with transformation into canonical forms for the marine-turbine power unit: (a) Convergence of the state variables x_2, x_4 and x_5 (blue lines) to the reference setpoints (red lines) and (b) control inputs u_1, u_2 applied to the marine-power generation system

2.1.9.5   Results on flatness-based control in successive loops

The obtained results about flatness-based control implemented in successive loops in the model of the marine power unit are given in Fig. 2.17 to Fig. 2.21, and confirm that this method performs fine. Unlike flatness-based control in successive loops and backstepping-type control the previously given optimal control method does not require the state-space description of the system to be found in a specific form. About flatness-based control in successive loops and about backstepping-type control it is known that they can not be directly applied to a dynamical system if the related state-space model is not found in the triangular (backstepping integral) form.

2.1.9.6   Results on fuzzy multiple-models-based control

The obtained results about fuzzy multi-model-based control in the model of the marine power unit are given in Fig. 2.27 to Fig. 2.31, and confirm that this method performs fine. Unlike multiple local models-based fuzzy control, the previously given nonlinear optimal control method uses only one linearization point and needs the solution of only one Riccati equation so as to compute the stabilizing feedback gains of the controller. Consequently, in terms of computation load, the proposed nonlinear optimal control method for the marine-turbine and synchronous generator power system is much more efficient.

2.1.9.7   Results on PID control

The obtained results about PID control in the model of the marine power unit are given in Fig. 2.32 to Fig. 2.36. The selection of the gains of the PID control can be performed with the Ziegler-Nichols technique. Unlike PID control, the previously given nonlinear optimal control method is of proven global stability, the selection of the controller parameters in it does not rely on a heuristic tuning procedure, and the stability of the control loop is assured in the case of changes of operating points.

2.2   FAULT DIAGNOSIS OF THE MARINE-TURBINE AND SYNCHRONOUS GENERATOR UNIT

2.2.1   OUTLINE

At a second stage, the chapter treats the problem of fault diagnosis for a marine-turbine and synchronous-generator power unit. The rapid growth of energy consumption worldwide, as well as the shift to green energy and the associated Net-zero objective have fostered the deployment of the use of renewable energy sources [205], [44], [198]. In this context, marine power generation units have a raising contribution to the energy balance of several countries where intensive tidal phenomena and strong sea currents appear [259], [249], [281]. The functioning of marine power generation units often takes place under harsh conditions which results into turbine or generator failures [256], [253], [280], [87]. Faults can emerge either at the mechanical or at the electrical part of marine power generators and may result into power shortages, costly restoration procedures or even complete damage of equipment [242], [207], [258]. Consequently, early fault diagnosis in marine-turbine power generators is important for avoiding critical conditions and for assuring the safe functioning of these power generation units and uninterrupted power supply [130], [69], [279].

Condition monitoring methods for marine current power systems may be developed either in the time domain or in the frequency domain [18], [189]. In the present chapter, the problem of fault diagnosis for marine-turbine power generation units is treated in time-domain with the use of a flatness-based disturbance observer [195], [188]. First, it is proven that the dynamic model of the marine-turbine power generation unit is differentially flat. The differential flatness property signifies that using state variable transformations (diffeomorphisms) the state-space description of the system can be written in the input-output linearized form and subsequently in the canonical Brunovsky form. For the latter state-space representation, the stabilization and control problem can be solved with the pole-placement technique, while the filtering and state-estimation problem can be treated with the use of the Kalman Filter's recursion. Actually, the implementation of Kalman Filter for the linearized equivalent description of the system, being followed by inverse state variable transformations allows for estimating the state variables of the initial nonlinear dynamics of the marine power generation unit, and this technique is known as Derivative-free nonlinear Kalman Filter [195].

At a next stage, by redesigning the previously noted Kalman Filtering method as a disturbance observer, simultaneous estimation of the nonmeasurable state vector elements of the marine power unit can be achieved as well as estimation of disturbance inputs that affect this system. Actually, the state vector of the marine power unit is extended by including as state variables in it the additive disturbance terms that affect the system and their time-derivatives. For the extended state-space description of the marine power unit the Kalman Filter's recursion provides estimates of the disturbance inputs. The identified cumulative terms of disturbances may represent model uncertainty, parametric changes and additive perturbations of the control inputs. By knowing the aggregate perturbations that affect the marine power system, the condition monitoring problem of the power unit is treated. The Kalman Filter-based disturbance estimator can identify in real-time even small disturbance signals, thus accomplishing the objective of incipient fault diagnosis. Moreover, by distinguishing between disturbances that affect the turbine part of the marine power unit and disturbances that affect the synchronous generator part of this power system one can also achieve fault isolation.

There has been much progress on methods for the reliable functioning of marine-turbine power generation units and the present chapter's fault diagnosis approach is a new contribution in this domain [136], [260], [252], [106], [183]. The problem of fault diagnosis of complex electric power generation systems is nontrivial because the following two objectives have to be achieved: (i) early detection of incipient parametric changes and (ii) fault isolation. The present chapter addresses both of these objectives. By using the differential flatness-properties of the marine power unit and by considering the additive disturbance inputs that affect this system as additional state variables, the state-space description of the power unit is transformed into the canonical Brunovsky form, which is both controllable and observable. For the extended state-space model of the system, the Kalman Filter is redesigned as a disturbance observer thus enabling to estimate in real-time the non-measurable state variables of the power unit and to detect with high precision additive disturbances even if the latter signals are of small magnitude. The Kalman Filter-based disturbance observer can provide separately estimates of perturbations that affect the mechanical part of the power unit (that is the turbine) and the electrical part of this power system (that is the synchronous generator), thus also allowing for fault isolation.

2.2.2 DIFFERENTIAL FLATNESS FOR MARINE-TURBINE AND SG POWER UNITS

As shown in the previous section, by successively differentiating the flat outputs $x_1 = \theta_t$ and $x_3 = \theta_g$ the marine-turbine and synchronous generator power unit is written in the following input-output linearized form

$$\ddot{x}_1 = f_1(x) + g_{11}(x)u_1 + g_{21}(x)u_2 \tag{2.149}$$

where about functions $f_1(x)$, $g_{11}(x)$, $g_{21}(x)$ one has

$$f_1(x) = \frac{1}{J_t}[-K_1(x_1 - x_3) - Dx_2] \tag{2.150}$$

$$g_{11}(x) = \frac{T_m}{J_t} \quad g_{21}(x) = 0 \tag{2.151}$$

$$x_3^{(3)} = f_2(x) + g_{12}(x)u_1 + g_{22}(x)u_2 \tag{2.152}$$

while about functions $f_2(x)$, $g_{12}(x)$, $g_{22}(x)$ one has

$$\begin{aligned}
f_2(x) &= \frac{1}{J_g}K_1(x_2 - x_4)\frac{n_g}{n_t} \\
&- \frac{D_g}{J_g}\{\frac{1}{J_g}[K_1(x_1 - x_3)\frac{n_g}{n_t} - D_gx_4 - \omega_0\frac{V_sx_5sin(x_3)}{x'_{d\Sigma}}]\} - \\
&- \omega_0\frac{V_ssin(x_3)}{x'_{d\Sigma}}\{-\frac{1}{T'_d}x_5 + \frac{1}{T_{do}}\frac{x_d-x'_d}{x_{d\Sigma}}V_scos(x_3)\} - \\
&\omega_0\frac{V_sx_5cos(x_3)x_4}{x'_{d\Sigma}}
\end{aligned} \tag{2.153}$$

$$g_{12}(x) = 0 \quad g_{22}(x) = -\omega_0\frac{V_ssin(x_3)}{x'_{d\Sigma}}\frac{1}{T_{do}} \tag{2.154}$$

Thus, one arrives at the following input-output linearized form for the system of the marine-turbine and synchronous generator

$$\begin{pmatrix} \ddot{x}_1 \\ x_3^{(3)} \end{pmatrix} = \begin{pmatrix} f_1(x) \\ f_2(x) \end{pmatrix} + \begin{pmatrix} g_{11}(x) & 0 \\ 0 & g_{22}(x) \end{pmatrix} \begin{pmatrix} u_1 \\ u_2 \end{pmatrix} \qquad (2.155)$$

Next, a stabilizing feedback controller for the marine-turbine and synchronous generator power unit is designed. The input-output linearized description of the power unit is rewritten after including additive disturbance terms in it. Thus one has

$$\ddot{x}_1 = f_1(x) + g_{11}(x)u_1 + g_{21}(x)u_2 + \tilde{d}_1$$
$$x_3^{(3)} = f_2(x) + g_{12}(x)u_1 + g_{22}(x)u_2 + \tilde{d}_2 \qquad (2.156)$$

The following virtual control inputs are defined: $v_1 = f_1(x) + g_{11}(x)u_1 + g_{21}(x)u_2$ and $v_2 = f_2(x) + g_{12}(x)u_1 + g_{22}(x)u_2$, where $\tilde{d}_1$ and $\tilde{d}_2$ denote cumulative disturbance terms. These disturbance inputs can be due to parametric changes and model uncertainty (multiplicative disturbances) or may represent control input perturbations (additive disturbances). The control inputs which are finally applied to the initial nonlinear model of the marine-turbine power generation unit are

$$\begin{pmatrix} u_1 \\ u_2 \end{pmatrix} = \begin{pmatrix} g_{11}(x) & g_{21}(x) \\ g_{12}(x) & g_{22}(x) \end{pmatrix}^{-1} \begin{pmatrix} v_1 - f_1(x) - \hat{\tilde{d}}_1 \\ v_2 - f_2(x) - \hat{\tilde{d}}_2 \end{pmatrix} \qquad (2.157)$$

2.2.3 DESIGN OF A KALMAN FILTER-BASED DISTURBANCE OBSERVER

The disturbances $\tilde{d}_1$ and $\tilde{d}_2$ are denoted as additional state variables. Thus, the state variables of the system become: $z_1 = x_1$, $z_2 = \dot{x}_1$, $z_3 = x_2$, $z_4 = \dot{x}_2$, $z_5 = \ddot{x}_2$, $z_6 = \tilde{d}_1$, $z_7 = \dot{\tilde{d}}_1$, $z_8 = \tilde{d}_2$, $z_9 = \dot{\tilde{d}}_2$. By assuming that the second-order time-derivatives of the additive disturbances are given by $\ddot{\tilde{d}}_1 = f_{\tilde{d}_1}$ and $\ddot{\tilde{d}}_2 = f_{\tilde{d}_2}$, the extended state-space description of the system is written as

$$\begin{pmatrix} \dot{z}_1 \\ \dot{z}_2 \\ \dot{z}_3 \\ \dot{z}_4 \\ \dot{z}_5 \\ \dot{z}_6 \\ \dot{z}_7 \\ \dot{z}_8 \\ \dot{z}_9 \end{pmatrix} = \begin{pmatrix} 0 & 1 & 0 & 0 & 0 & 0 & 0 & 0 & 0 \\ 0 & 0 & 0 & 0 & 0 & 1 & 0 & 0 & 0 \\ 0 & 0 & 0 & 1 & 0 & 0 & 0 & 0 & 0 \\ 0 & 0 & 0 & 0 & 1 & 0 & 0 & 0 & 0 \\ 0 & 0 & 0 & 0 & 0 & 0 & 0 & 1 & 0 \\ 0 & 0 & 0 & 0 & 0 & 0 & 1 & 0 & 0 \\ 0 & 0 & 0 & 0 & 0 & 0 & 0 & 0 & 0 \\ 0 & 0 & 0 & 0 & 0 & 0 & 0 & 0 & 1 \\ 0 & 0 & 0 & 0 & 0 & 0 & 0 & 0 & 0 \end{pmatrix} \begin{pmatrix} z_1 \\ z_2 \\ z_3 \\ z_4 \\ z_5 \\ z_6 \\ z_7 \\ z_8 \\ z_9 \end{pmatrix} + \begin{pmatrix} 0 & 0 & 0 & 0 \\ 1 & 0 & 0 & 0 \\ 0 & 0 & 0 & 0 \\ 0 & 0 & 0 & 0 \\ 0 & 1 & 0 & 0 \\ 0 & 0 & 0 & 0 \\ 0 & 0 & 1 & 0 \\ 0 & 0 & 0 & 0 \\ 0 & 0 & 0 & 1 \end{pmatrix} \begin{pmatrix} v_1 \\ v_2 \\ f_{\tilde{d}_1} \\ f_{\tilde{d}_2} \end{pmatrix}$$
$$\qquad (2.158)$$

Denoting the state vector $Z = [z_1, z_2, z_3, z_4, z_5, z_5, z_6, z_7, z_8, z_9]^T$, the associated outputs of the system are z_1 and z_3, thus the measurement equation of the system becomes

$$\begin{pmatrix} z_1^m \\ z_2^m \end{pmatrix} = \begin{pmatrix} 1 & 0 & 0 & 0 & 0 & 0 & 0 & 0 & 0 \\ 0 & 0 & 1 & 0 & 0 & 0 & 0 & 0 & 0 \end{pmatrix} Z \qquad (2.159)$$

or equivalently the extended state-space description of the power unit is written as

$$
\begin{aligned}
\dot{Z} &= A_e Z + B_e v \\
Z_e^m &= C_e Z
\end{aligned}
\tag{2.160}
$$

The stabilizing feedback control for the marine-turbine and synchronous generator power unit is given by

$$
\begin{aligned}
v_1 &= \ddot{z}_{1,d} - k_{11}(\dot{z}_1 - \dot{z}_{1,d}) - k_{21}(z_1 - z_{1,d}) - \hat{\tilde{d}}_1 \\
v_2 &= z_{2,d}^{(3)} - k_{12}(\ddot{z}_3 - \ddot{z}_{3,d}) - k_{22}(\dot{z}_3 - \dot{z}_{3,d}) - k_{32}(z_3 - z_{3,d}) - \hat{\tilde{d}}_2
\end{aligned}
\tag{2.161}
$$

The extended state-space model is observable. To obtain estimates of the additive disturbance terms $\hat{\tilde{d}}_1$ and $\hat{\tilde{d}}_2$ the following disturbance observer is used

$$
\dot{Z} = A_{e,o} Z + B_{e,o} v + K_f (Z_e^m - \hat{Z}_e^m)
\tag{2.162}
$$

where the matrices which appear in the observer's equation are $A_{e,o} = A_e$, $C_{e,o} = C_e$ and

$$
B_{e,o}^T = \begin{pmatrix} 0 & 1 & 0 & 0 & 0 & 0 & 0 & 0 & 0 \\ 0 & 0 & 0 & 0 & 1 & 0 & 0 & 0 & 0 \end{pmatrix}
\tag{2.163}
$$

while the disturbance observer's gain matrix K_f is given by Kalman Filter's recursion [198]. The application of Kalman Filtering on the linearized equivalent description of the power system is the so-called *Derivative-free nonlinear Kalman Filter*. Matrices, $A_{e,o}$, $B_{e,o}$ and $C_{e,o}$ are discretized using common discretization methods. This provides matrices A_d, B_d and C_d. Matrices Q and R denote the process and measurement noise covariance matrices. Matrix P is the state-vector error covariance matrix. The Kalman Filter comprises a *measurement update* and a *time update* stage [195].

measurement update:

$$
\begin{aligned}
K_f(k) &= P^-(k) C_d^T [C_d P^-(k) C_d^T + R]^{-1} \\
\hat{x}(k) &= \hat{x}^-(k) + K_f(k)[z_m - \hat{z}_m] \\
P(k) &= P^-(k) - K_f(k) C_d P^-(k)
\end{aligned}
\tag{2.164}
$$

time update:

$$
\begin{aligned}
P^-(k+1) &= A_d P(k) A_d^T + Q \\
\hat{x}^-(k+1) &= A_d \hat{x}(k) + B_d v(k)
\end{aligned}
\tag{2.165}
$$

2.2.4 STATISTICAL TESTS FOR FAULT DETECTION AND ISOLATION

2.2.4.1 Fault detection

The residuals' sequence, that is the differences between (i) the zero-valued disturbance inputs $\tilde{d}_i$ $i = 1, 2$ of the marine-turbine power unit in the fault-free case and

(ii) the estimated values of the disturbance inputs $\hat{d}_i$ $i = 1, 2$ which are provided by the Kalman Filter-based disturbance observer (Fig. 2.37) is a discrete error process e_k with dimension $m \times 1$ (here $m = N$ is the dimension of the output measurements vector). Actually, it is a zero-mean Gaussian white-noise process with covariance given by E_k.

A conclusion can be stated, based on a measure of certainty that the marine-turbine power generation unit has not been subjected to a fault. To this end, the following *normalized error square* (NES) is defined [195]

$$\varepsilon_k = e_k^T E_k^{-1} e_k \qquad (2.166)$$

The sum of this normalized residuals' square follows a χ^2 distribution, with a number of degrees of freedom that is equal to the dimension of the residuals' vector. The form of the χ^2 distribution for various degrees of freedom is shown in Fig. 2.38. An appropriate test for the normalized error sum is to numerically show that the following condition is met within a level of confidence (according to the properties of the χ^2 distribution)

$$E\{\varepsilon_k\} = m \qquad (2.167)$$

This can be achieved using statistical hypothesis tests, which are associated with confidence intervals. For instance, a 95% confidence interval is frequently applied, which is specified using the probability region $100(1 - a)$ with $a = 0.05$. Actually, a two-sided probability region is considered cutting-off two end tails of 2.5% each. For M runs, the normalized error square that is obtained is given by

$$\bar{\varepsilon}_k = \frac{1}{M} \sum_{i=1}^{M} \varepsilon_k(i) = \frac{1}{M} \sum_{i=1}^{M} e_k^T(i) E_k^{-1}(i) e_k(i) \qquad (2.168)$$

where ε_i stands for the i-th run at time t_k. Then $M\bar{\varepsilon}_k$ will follow a χ^2 density with Mm degrees of freedom. This condition can be checked using a χ^2 test. The hypothesis holds, if the following condition is satisfied

$$\bar{\varepsilon}_k \in [\zeta_1, \zeta_2] \qquad (2.169)$$

where ζ_1 and ζ_2 are derived from the tail probabilities of the χ^2 density.

2.2.4.2   Fault isolation

By applying the statistical test into the individual components of the marine-turbine and synchronous generator power unit, it is also possible to find out the specific component that has been subjected to a fault [195], [198]. For a marine-turbine power unit of n parameters suspected for change, one has to carry out n χ^2 statistical change detection tests, where each test is applied to the subset that comprises parameters $i - 1$, i and $i + 1, i = 1, 2, \cdots, n$. Actually, out of the n χ^2 statistical change detection tests, the ones that exhibit the highest score are those that identify the parameter that has been subject to change.

In the case of multiple faults, one can identify the subset of parameters that has been subjected to change by applying the χ^2 statistical change detection test according to a combinatorial sequence. This means that

$$\binom{n}{k} = \frac{n!}{k!(n-k)!} \tag{2.170}$$

tests have to take place, for all clusters in the marine-turbine power unit, that finally comprise n, $n-1$, $n-2$, $\cdots$, 2, 1 parameters. Again the χ^2 tests that give the highest scores indicate the parameters which are most likely to have been subjected to change.

As previously noted, the concept of the proposed statistical fault detection and isolation test is simple. The residuals' vector is formed by taking at each sampling instance the differences between the estimated values of the additive disturbance inputs $[\hat{d}_1, \hat{d}_2]$ and the zero-valued vector $[\tilde{d}_1 = 0, \tilde{d}_2 = 0]$ that is the values of the disturbance inputs in the fault-free case. The sum of the squares of the residuals' vector, weighted by the inverse of the residuals' covariance matrix, stands for a stochastic variable which follows the χ^2 distribution. Actually, this is a multi-dimensional χ^2 distribution and the number of its degrees of freedom is equal to the dimension of the residuals' vector. Since, there are 2 estimated disturbance inputs at the marine-turbine power unit, the residuals' vector is of dimension 2 and the number of degrees of freedom is also 2. Next, from the properties of the χ^2 distribution, the mean value of the aforementioned stochastic variable in the fault-free case should be also 2. However, due to small fluctuations of the estimated variables, the value of the statistical test in the fault-free case will not be precisely equal to 2 but it may vary within a small range around this value. This range is determined by the confidence intervals of the χ^2 distribution. For a probability of 98% to get a value of the stochastic variable about 2, the associated confidence interval is given by the lower bound $L = 1.84$ and by the upper bound $U = 2.16$. Consequently, as long as the statistical test provides an indication that the aforementioned stochastic variable is in the interval $[L, U]$, the functioning of the power unit can be concluded to be free of faults. On the other side, when the bounds of the previously given confidence interval are exceeded it can be concluded that the power unit has been subjected to a fault. Finally, by performing the statistical test in subspaces of the power unit's state-space model, where each subspace is associated with different components, one can also achieve fault isolation. This signifies that the specific component that has caused the malfunctioning of the power unit can be identified.

2.2.5 SIMULATION TESTS

2.2.5.1 Performance of the Kalman Filter-based disturbance estimator

Indicative results about flatness-based control with transformation of the model of the marine power unit in canonical (Brunovsky) form are given first in Fig. 2.39.

Besides, results about disturbances' estimation in the dynamic model of the marine power unit with the use of a Kalman Filter-based disturbance observer are given in Fig. 2.40 to Fig. 2.44. The real values of the disturbance signals are depicted in blue color while their estimates, which have been provided by the Kalman Filter, are shown in green color. The simulation tests demonstrate that the disturbance observer allows for simultaneous estimation of the non-measurable state variables and of the additive disturbance inputs that were affecting the marine-turbine power unit.

The parameters of the simulation experiments are given below:

Test (setpoint) 1: Total simulation time $T = 40$sec. Setpoints for the state variables which are also flat outputs of the system: $x_{1,d}(t) = 0.1 + 0.01t$ and $x_{3,d}(t) = 0.1 + 0.01t \cdot \frac{n_L}{n_g}$ for $t \in [0, T]$. Constant disturbance signals $d_1(t) = 0.1$ and $d_2(t) = 0.1$ for $t \in [0, T]$. Sinusoidal disturbance signals: $d_1(t) = 0.1 + 0.2sin(\frac{2\pi t}{T})$ and $d_2(t) = 0.1 + 0.2sin(2\frac{\pi t}{T})$ for $t \in [0, T]$.

Test (setpoint) 2: Total simulation time $T = 40$sec. Setpoints for the state variables which are also flat outputs of the system: (i) $x_{1,d}(t) = 0.1 + 0.01t$ and $x_{3,d}(t) = 0.1 + 0.01t \cdot \frac{n_L}{n_g}$ for $t \in [0, \frac{T}{2})$ and (ii) $x_{1,d} = 0.1 + 0.01\frac{T}{2} + 0.10(t - \frac{T}{2})$ and $x_{3,d}(t) = 0.1 + 0.01\frac{T}{2}\frac{n_L}{n_g} + 0.10(t - \frac{T}{2})\frac{n_L}{n_g}$ for $t \in [\frac{T}{2}, T]$. Constant disturbance signals (i) $d_1(t) = 0.1$ and $d_2(t) = 0.2$ for $t \in [0, \frac{T}{2}$ for $t \in [0, \frac{T}{2}]$ and (ii) $d_1(t) = 0.2$ and $d_2(t) = 0.3$ for $t \in [\frac{T}{2}, T]$. Sinusoidal disturbance signals: (i) $d_1(t) = 0.1 + 0.05cos(\frac{2\pi t}{0.75T})$ and $d_2(t) = 0.1 + 0.05cos(\frac{2\pi t}{0.75T})$ for $t \in [0, \frac{T}{2})$ and (ii) $d_1(t) = 0.2 + 0.05cos(\frac{2\pi t}{0.5T})$ and $d_2(t) = 0.2 + 0.05cos(\frac{2\pi t}{0.5T})$ for $t \in [\frac{T}{2}, T]$.

Test (setpoint) 3: Total simulation time $T = 40$sec. Setpoints for the state variables which are also flat outputs of the system: (i) $x_{1,d}(t) = 0.1 + 0.02t$ and $x_{3,d}(t) = 0.1 + 0.02t \cdot \frac{n_L}{n_g}$ for $t \in [0, \frac{T}{2})$ and (ii) $x_{1,d} = 0.05 + 0.02\frac{T}{2} - 0.10(t - \frac{T}{2})$ and $x_{3,d}(t) = 0.05 + 0.02\frac{T}{2}\frac{n_L}{n_g} - 0.0003(t - \frac{T}{2})\frac{n_L}{n_g}$ for $t \in [\frac{T}{2}, T]$. Constant disturbance signals (i) $d_1(t) = 0.2$ and $d_2(t) = 0.3$ for $t \in [0, \frac{T}{2}$ for $t \in [0, \frac{T}{2}]$ and (ii) $d_1(t) = 0.1$ and $d_2(t) = 0.2$ for $t \in [\frac{T}{2}, T]$. Sinusoidal disturbance signals: (i) $d_1(t) = 0.05 + 0.05sin(\frac{2\pi t}{T}) + 0.05sin(\frac{2\pi t}{0.75T})$ and $d_2(t) = 0.05 + 0.05cos(2\frac{\pi t}{T}) + 0.05cos(\frac{2\pi t}{0.75T})$ for $t \in [0, \frac{T}{2})$ and (ii) $d_1(t) = 0.05 + 0.05sin(\frac{2\pi T}{T}) + 0.05sin(\frac{2\pi T}{0.75T}) + 0.05sin(\frac{2\pi T}{0.5T})$ and $d_2(t) = 0.05 + 0.05cos(2\frac{\pi t}{T}) + 0.05cos(\frac{2\pi t}{0.75T}) + 0.05cos(\frac{2\pi t}{0.5T})$ for $t \in [\frac{T}{2}, T]$.

Test (setpoint) 4: Total simulation time $T = 40$sec. Setpoints for the state variables which are also flat outputs of the system: (i) $x_{1,d}(t) = 0.1 + 0.01t$ and $x_{3,d}(t) = 0.01 + 0.01 \cdot \frac{n_L}{n_g}$ for $t \in [0, \frac{T}{3})$, (ii) $x_{1,d}(t) = 0.1 + 0.01\frac{T}{3} + 0.10(t - \frac{T}{3})$ and $x_{3,d}(t) = 0.01 + 0.01\frac{T}{3}\frac{n_L}{n_g} + 0.015(t - \frac{T}{3})\frac{n_L}{n_g}$ for $t \in [\frac{T}{3}, 2\frac{T}{3})$ and (iii) $x_{1,d}(t) = 0.1 + 0.01\frac{T}{3} + 0.10\frac{T}{3} + 0.18(t - \frac{2T}{3})$ and $x_{3,d}(t) = 0.1 + 0.01\frac{T}{3}\frac{n_L}{n_g} + 0.015\frac{T}{3}\frac{n_L}{n_g} + 0.02(t - \frac{2T}{3})\frac{n_L}{n_g}$ for $t \in [\frac{2T}{3}, T]$. Constant disturbance signals (i) $d_1(t) = 0.1$ and $d_2(t) = 0.1$ for

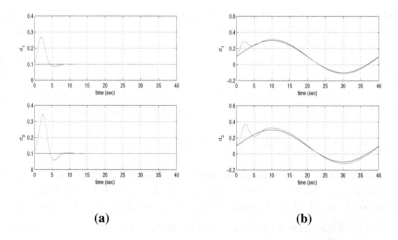

(a)   (b)

Figure 2.40 KF-based estimation (green lines) of disturbance inputs $\tilde{d}_1$ and $\tilde{d}_2$ (blue lines) applied respectively to the turbine and the generator when tracking setpoint 1: (a) piecewise constant disturbance inputs and (b) time-varying (sinusoidal) disturbance inputs

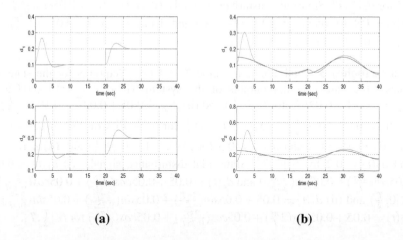

(a)   (b)

Figure 2.41 KF-based estimation (green lines) of disturbance inputs $\tilde{d}_1$ and $\tilde{d}_2$ (blue lines) applied respectively to the turbine and the generator when tracking setpoint 2: (a) piecewise constant disturbance inputs and (b) time-varying (sinusoidal) disturbance inputs

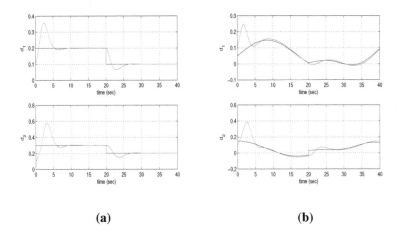

(a) **(b)**

Figure 2.42 KF-based estimation (green lines) of disturbance inputs $\tilde{d}_1$ and $\tilde{d}_2$ (blue lines) applied respectively to the turbine and the generator when tracking setpoint 3: (a) piecewise constant disturbance inputs and (b) time-varying (sinusoidal) disturbance inputs

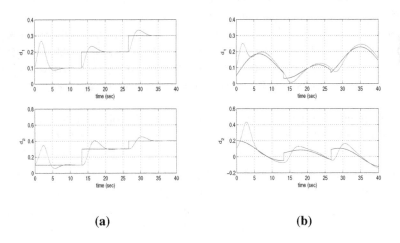

(a) **(b)**

Figure 2.43 KF-based estimation (green lines) of disturbance inputs $\tilde{d}_1$ and $\tilde{d}_2$ (blue lines) applied respectively to the turbine and the generator when tracking setpoint 4: (a) piecewise constant disturbance inputs and (b) time-varying (sinusoidal) disturbance inputs

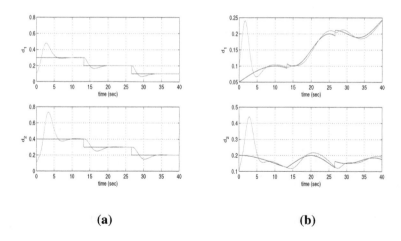

(a)   (b)

Figure 2.44 KF-based estimation (green lines) of disturbance inputs $\tilde{d}_1$ and $\tilde{d}_2$ (blue lines) applied respectively to the turbine and the generator when tracking setpoint 5: (a) piecewise constant disturbance inputs and (b) time-varying (sinusoidal) disturbance inputs

$t \in [0, \frac{T}{3})$, (ii) $d_1(t) = 0.2$ and $d_2(t) = 0.3$ for $t \in [\frac{T}{3}, \frac{2T}{3})$ and (iii) $d_1(t) = 0.3$ and $d_2(t) = 0.4$ for $t \in [2\frac{T}{3}, T]$. Sinusoidal disturbances: (i) $d_1(t) = 0.05 + 0.05sin(\frac{2\pi t}{T}) + 0.05sin(\frac{2\pi t}{0.75T}) + 0.05sin(\frac{2\pi t}{0.5T})$ and $d_2(t) = 0.05 + 0.05cos(\frac{2\pi t}{T}) + 0.05cos(\frac{2\pi t}{0.75T}) + 0.05cos(\frac{2\pi t}{0.5T})$ for $t \in [0, \frac{T}{3})$, (ii) $d_1(t) = 0.05 + 0.05sin(\frac{2\pi t}{T}) - 0.05sin(\frac{2\pi t}{0.75T}) + 0.05sin(\frac{2\pi t}{0.5T})$ and $d_2(t) = 0.05 + 0.05cos(\frac{2\pi t}{T}) - 0.05cos(\frac{2\pi t}{0.75T}) + 0.05cos(\frac{2\pi t}{0.5T})$ for $t \in [\frac{T}{3}, \frac{2T}{3})$ and (iii) $d_1(t) = 0.10 - 0.05sin(\frac{2\pi t}{T}) + 0.05sin(\frac{2\pi t}{0.75T}) - 0.05sin(\frac{2\pi t}{0.5T})$ and $d_2(t) = d_2(t) = -0.05cos(\frac{2\pi t}{T}) + 0.05cos(\frac{2\pi t}{0.75T}) - 0.05cos(\frac{2\pi t}{0.5T})$ for $t \in [\frac{2T}{3}, T]$.

Test (setpoint) 5: Total simulation time $T = 40sec$. Setpoints for the state variables which are also flat outputs of the system: (i) $x_{1,d}(t) = 0.1 + 0.01t$ and $x_{3,d}(t) = 0.01 + 0.01 \cdot \frac{n_t}{n_g}$ for $t \in [0, \frac{T}{3})$, (ii) $x_{1,d}(t) = 0.1 + 0.01\frac{T}{3} + 0.10(t - \frac{T}{3})$ and $x_{3,d}(t) = 0.01 + 0.01\frac{T}{3}\frac{n_t}{n_g} + 0.10(t - \frac{T}{3})\frac{n_t}{n_g}$ for $t \in [\frac{T}{3}, \frac{2T}{3})$ and (iii) $x_{1,d}(t) = 0.1 + 0.01\frac{T}{3} + 0.10\frac{T}{3} - 0.10(t - \frac{2T}{3})$ and $x_{3,d}(t) = 0.1 + 0.01\frac{T}{3}\frac{n_t}{n_g} + 0.10\frac{T}{3}\frac{n_t}{n_g} - 0.0003(t - \frac{2T}{3})\frac{n_t}{n_g}$ for $t \in [\frac{2T}{3}, T]$. Constant disturbance signals (i) $d_1(t) = 0.3$ and $d_2(t) = 0.4$ for $t \in [0, \frac{T}{3})$, (ii) $d_1(t) = 0.2$ and $d_2(t) = 0.3$ for $t \in [\frac{T}{3}, \frac{2T}{3})$ and (iii) $d_1(t) = 0.1$ and $d_2(t) = 0.2$ for $t \in [2\frac{T}{3}, T]$. Sinusoidal disturbance signals: (i) $d_1(t) = 0.05 + 0.05sin(\frac{2\pi t}{T})$ and $d_2(t) = 0.05 + 0.05cos(\frac{2\pi t}{T})$ for $t \in [0, \frac{T}{3})$ and (ii) $d_1(t) = 0.05 + 0.05sin(\frac{2\pi t}{0.5T})$ and $d_2(t) = 0.05 + 0.05cos(\frac{2\pi t}{0.5T})$ for $t \in [\frac{T}{3}, \frac{2T}{3})$ and (iii) $d_1(t) = 0.05 + 0.05sin(\frac{2\pi t}{0.75T}) + 0.05sin(\frac{2\pi t}{0.5T})$ and $d_2(t) = 0.05 + 0.05cos(\frac{2\pi t}{0.75T}) + 0.05cos(\frac{2\pi t}{0.5T})$ for $t \in [\frac{2T}{3}, T]$.

2.2.5.2 Statistical tests for fault diagnosis

Considering $n = 2$ disturbance inputs for the marine-turbine power generation unit, to conclude the normal functioning of the power unit, the previously analyzed statistical test should take a value that is very close to the mean value of the χ^2 distribution, that is 2.

The residuals' vector is formed by taking at each sampling instance the differences between the estimated values of the additive disturbance inputs $[\hat{\tilde{d}}_1, \hat{\tilde{d}}_2]$ and the zero-valued vector $[\tilde{d}_1 = 0, \tilde{d}_2 = 0]$, that is the values of the disturbance inputs in the fault-free case. Thus, the dimension of the residuals' vector is $m = 2$. Considering that the number of residual vector samples is $M = 2000$ and using a 98% confidence interval for the χ^2 distribution the fault thresholds can be defined as $L = 1.84$ and $U = 2.16$. In an equivalent manner, when the statistical test is applied exclusively to the marine-turbine or exclusively to the synchronous generator, the fault thresholds are defined as $L = 0.89$ and $U = 1.11$. Moreover, when the statistical test is applied only to the synchronous generator the fault thresholds are defined again as $L = 0.89$ and $U = 1.11$.

In case of fault detection with χ^2 tests, the Derivative-free nonlinear Kalman Filter has been used as a disturbance estimator and the statistical χ^2 test has been applied to the power system that comprises both the marine turbine and the synchronous generator. In particular, the following cases have been examined:

- case (0): functioning of the marine-turbine and synchronous generator power unit in the fault-free condition. The associated results are depicted in Fig. 2.45.
- case (1) functioning of the power generation unit under additive input disturbance at both the turbine and the generator. The associated results are depicted in Fig. 2.46.
- case (2): functioning of the power generation unit under additive input disturbance at the turbine. The associated results are depicted in Fig. 2.47.
- case (3) functioning of the power generation unit under additive input disturbance at the generator. The associated results are depicted in Fig. 2.48.

It has been confirmed that in all aforementioned cases the proposed condition monitoring method was capable of detecting incipient faults appearing in components of the power unit

The proposed fault diagnosis approach for marine-turbine power generation units is novel. Usually, observer-based fault diagnosis methods are based on the processing of the residuals of the estimated outputs of the monitored system, while the definition of fault thresholds in them is often performed in an empirical manner. On the

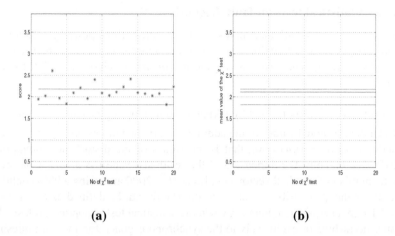

(a)   **(b)**

Figure 2.45   Test case 0 when neither a fault exists at the marine turbine nor a fault has taken place at the synchronous generator: (a) values of successive χ^2 tests performed at the integrated power system and (b) mean value of the χ^2 tests for the integrated power system

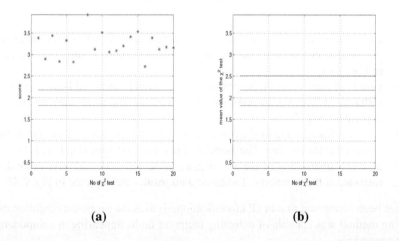

(a)   **(b)**

Figure 2.46   Test case 1 when a fault exists at the marine turbine and a fault has taken place also at the synchronous generator: (a) values of successive χ^2 tests performed at the integrated power system and (b) mean value of the χ^2 tests for the integrated power system

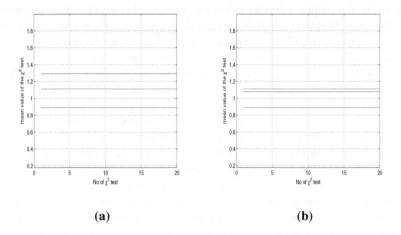

(a)   **(b)**

Figure 2.47   Test case 2 when a fault exists at the marine turbine and no fault has taken place at the synchronous generator: (a) mean value of the χ^2 tests performed at the marine turbine and (b) mean value of the χ^2 tests performed at the synchronous generator

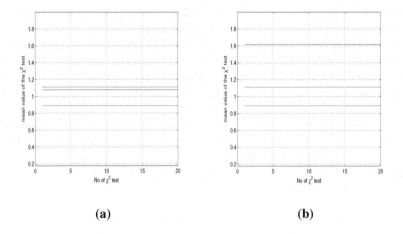

(a)   **(b)**

Figure 2.48   Test case 3 when no fault exists at the marine turbine and a fault has taken place at the synchronous generator: (a) mean value of the χ^2 tests performed at the marine turbine and (b) mean value of the χ^2 tests performed at the synchronous generator

other side, the chapter's developments are innovative because (1) fault detection and isolation is performed through the statistical processing of the estimated disturbance inputs of the system, (2) fault thresholds are defined using the statistical properties of the residuals of the estimated disturbance inputs, thus allowing for precise monitoring of the evolution of component failures and for the early launching of alarms.

About (1), it is noted that state and disturbances estimation is performed with a novel and statistically optimal disturbance observer which allows for precise estimation of the non-measurable state variables of the system and of model uncertainty terms or exogenous perturbations that affect the system. Using the differential flatness properties of the monitored system, one achieves transformation of the system's state-space description into the input-output linearized form and also into the canonical Brunovsky form. In this new state-space representation, model uncertainty and disturbances appear as additive input perturbations. Moreover, by extending the state-vector of the system with the inclusion of the perturbation terms and of their time-derivatives as additional state variables one arrives at a new state-space model of higher dimension. Next, with the use of the Kalman Filter recursion one can solve optimally both the state vector and the disturbance vector estimation problem. About (2), it is known that the selection of fault thresholds in condition monitoring procedures is a non-trivial problem which in most cases is treated in an empirical manner. The precise definition of fault thresholds allows for the early detection of incipient failures and for the minimization of false alarms. In this chapter, to define fault thresholds the statistical processing of the residuals of the estimated additive disturbance inputs have been used. The residuals' sequence has been generated by subtracting at each sampling instance the estimated disturbances vector for the zero reference value. Next, one can use the stochastic variable (statistical test) of the sum of the squares of the residuals' vector weighted by the associated covariance matrix. This variable follows the χ^2 distribution with a number of degrees of freedom, which is equal to the dimension of the disturbances' vector. The confidence intervals of this distribution can be used for defining precise fault thresholds which demonstrate the existence of failures at a high confidence level of the order of 95% or more. When the above-noted statistical test persistently exceeds the confidence intervals, then one has a strong indication about the existence of malfunctioning. By repeating the statistical test in subspaces of the state-space model of the monitored system one can achieve also fault isolation.

Finally, it is noted that the proposed disturbance observer and the presented fault diagnosis approach can find use in more types of renewable energy systems, such as wind-power generation units, hydro-power generation units, biomass-fed gas-turbine power generators and microgrids consisting of renewable energy sources.

Additionally, the chapter's developments can be applied to condition monitoring of the power electronics that enable the connection of the previously listed power systems with the main electricity grid or with microgrids [204], [206]. By treating the fault diagnosis problem with the aforementioned approaches, the rate of failures or the risk for emergence of cascading events and critical conditions will be minimized. Moreover, there will be more effective maintenance and longer life-time of associated equipment.

3 Fault diagnosis for electricity microgrids and gas processing units

3.1 FAULT DIAGNOSIS FOR ELECTRIC POWER DC MICROGRIDS

3.1.1 OUTLINE

The chapter treats first the fault diagnosis problem for electric power DC microgrids. The use of microgrids is rapidly deploying as a solution to the p], [of uninterrupted power supply for civil and industrial use. Microgrids can supply with energy small and geographically isolated areas such as islands. By providing the microgrids' power to such territories, the cost for developing new infrastructure for the electric power and distribution system can be significantly waived. Furthermore microgrids can supply with energy industrial units (e.g. cement mills or steel rolling mills). By supplying electric power to industrial units through dedicated for this purpose microgrids, industrial production becomes less vulnerable to faults, perturbations and load variations in the main electricity grid [198], [203], [98], [100], [60], [114]. At the same time, due to the harsh operating conditions of these systems and due to exposure of their control software to malignant human intrusion, the development of efficient fault and cyber-attack diagnosis approaches becomes a necessity. By administering the fault and cyber-attack detection and isolation problems for such a type of power units, one can pursue the microgrids' optimized functioning.

As noted above, due to being exposed to variable and harsh operating conditions, DC microgrids undergo failures [2], [251], [179], [163]. Furthermore, when DC microgrids function as part of networked control schemes they become exposed to cyber-attacks targeting their control and data acquisition software. It is important to accomplish early fault and cyber-attack detection and incipient failure diagnosis for DC microgrids so as to avoid excessive damage of this equipment and to take action for its repair [123], [67], [120], [155] . Major objectives in fault and cyber-attack detection and isolation for DC microgrids, as well as in the diagnosis of cyber-attacks against the DC microgrids' control software, are (i) the early diagnosis of malfunctioning [107], [91], [154], (ii) the spotting of incipient parametric changes [21], [3], [177] , and (iii) the definition of reliable fault thresholds and decision criteria for the existence of failures [63], [14], [131], [28].

Fault diagnosis for dynamical systems is a non-trivial problem that has been treated in certain cases with the use of state observers and statistical filters [275], [272]. To achieve resilient operation of DC microgrids against equipment faults and

DOI: 10.1201/9781003527657-3

cyber-attacks, observer-based state estimation and condition monitoring methods have been developed. In [72] Luenberger observers together with a bank of unknown input observers are used for solving the state-estimation and the disturbance estimation problem in DC microgrids. In [250] an approach to the state estimation problem of DC microgrids is developed using H-infinity observers with gains which are computed through the solution of LMIs. In [12] the state-space model of a DC-microgrid is written in an LPV form and a sliding-mode observer is developed about it, in an aim to apply state estimation-based condition monitoring. In [233] an overview of the use of state-observers and of associated residual generation techniques is given and related applications to the cyber-attacks detection problem in DC microgrids are highlighted. In [215] the use of sliding-mode observers is proposed for state estimation-based condition monitoring in DC microgrids. In [41] the model of DC microgrids is transformed into an observable form and next state estimation and fault diagnosis is performed with the use of sliding-mode observers. Finally, in [166] the DC microgrids' model is written in an LPV form and based on this description, H_∞ disturbance observers are developed aiming at achieving state estimation and fault diagnosis. It is noteworthy that despite the extent of research on the use of observers for fault diagnosis in DC microgrids, the problem of fault threshold selection in such systems has not been sufficiently analyzed and is often treated in an empirical manner.

In the present chapter, a systematic method is developed for fault detection and isolation in DC micro-grids, as well as for the detection of cyber-attacks and malicious signals' interference in their control loop. The method relies on the state vector and outputs estimation for the DC micro-grids with the use of a robust filtering scheme, known as H-infinity Kalman Filter [195], [202]. Since the H-infinity Kalman Filter is primarily addressed to linear dynamical systems, to enable its use in the nonlinear model of DC microgrids, the state-space description of the DC microgrids undergoes linearization through Taylor series expansion [196], [197]. The linearization is performed around a temporary operating point, which is recomputed at each time step of the condition monitoring method [196-197]. Furthermore, the chapter's approach defines fault thresholds in a systematic and precise manner. The H-infinity Kalman Filter is considered to emulate the function of the DC microgrids in the fault-free and cyber-attack-free case. The filter's output is compared against the real output of the DC microgrids, thus generating the residual vectors' sequence [270], [268], [20].

Next, it is shown that the sum of the square of the residuals' vectors, weighted with the inverse of their covariance matrix stands for a stochastic variable which follows the χ^2 distribution [188], [18], [189]. Moreover, by using the confidence intervals of this distribution one can conclude with a high confidence level (for instance a certainty measure of the order of 95% or 98%) the normal functioning of the system or the appearance of a fault. As long as this stochastic variable falls between the upper and lower bounds of the confidence interval one can infer that the DC microgrid is in the fault-free mode. On the other side when the upper or lower bound is persistently exceeded, then the existence of a fault or cyber-attack can be diagnosed and an alarm can be launched. Furthermore, by applying the statistical test selectively to subspaces of the residuals' vector of the DC microgrids one can achieve also fault and cyber-attack isolation. .

3.1.2 DYNAMIC MODEL OF THE DC MICROGRIDS

3.1.2.1 Power system's dynamics

An indicative microgrid which can supply with power an industrial unit is shown in the following diagram (Fig. 3.1) and is described by the following set of equations [203], [98], [100]:

$$\dot{x}_1 = -\frac{1}{R_1 C_1} x_1 - \frac{1}{C_1} x_3 + \frac{1}{R_1 C_1} V_{PV} \tag{3.1}$$

$$\dot{x}_2 = -\frac{1}{R_2 C_2} x_2 - \frac{1}{C_2} x_3 - \frac{1}{C_2} x_3 u_1 + \frac{1}{R_2 C_2} x_9 \tag{3.2}$$

$$\dot{x}_3 = \frac{1}{L_3} (x_1 - x_2 - R_{o_1} x_3) + \frac{1}{L_3} (x_2 + (R_{o_1} - R_{o_2}) x_3) u_1 \tag{3.3}$$

$$\dot{x}_4 = -\frac{1}{R_4 C_4} x_4 - \frac{1}{C_4} x_6 + \frac{1}{R_4 C_4} V_B \tag{3.4}$$

$$\dot{x}_5 = -\frac{1}{R_5 C_5} x_5 + \frac{1}{C_5} x_6 + \frac{1}{R_5 C_5} x_9 - \frac{1}{C_5} x_6 u_2 \tag{3.5}$$

$$\dot{x}_6 = \frac{1}{L_6} x_4 - \frac{1}{L_6} x_5 - \frac{R_{o4}}{L_6} + \frac{1}{L_6} x_5 u_2 \tag{3.6}$$

$$\dot{x}_7 = -\frac{1}{R_7 C_7} x_7 + \frac{1}{C_7} x_8 + \frac{1}{R_7 C_7} x_9 \tag{3.7}$$

$$\dot{x}_8 = \frac{1}{L_8} V_s u_3 - \frac{R_{08}}{L_8} x_8 - \frac{1}{L_8} x_7 \tag{3.8}$$

$$\dot{x}_9 = \frac{1}{C_9} \left(\frac{x_2 - x_9}{R_2} + \frac{x_5 - x_9}{R_5} + \frac{x_7 - x_9}{R_7} - x_9 \frac{1}{R_L} \right) \tag{3.9}$$

The state variables of the dynamic model of the DC microgrid are defined as follows [203]: x_1 is voltage V_{c_1} at capacitor C_1, x_2 is voltage V_{c_2} at capacitor C_2, x_3 is current i_{L_3} at the inductor L_3, x_4 is voltage V_{C_4} at the capacitor C_4, x_5 is voltage V_{C_5} at the capacitor C_5, x_6 is current i_{L_6} at the inductor L_6, x_7 is voltage V_{C_7} at the capacitor C_7, x_8 is current i_{L_8} at the inductor L_8 and x_9 is voltage V_{C_9} at the capacitor C_9.

Control is implemented with the use of the Pulse Width Modulation (PWM) approach. The control inputs of the dynamic model of the DC microgrid are: u_1 which stands for the duty cycle of the switch at the PV circuit, u_2 which is the duty cycle

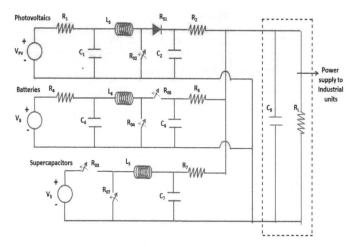

Figure 3.1 Use of a microgrid that comprises photovoltaics, batteries and supercapacitors to provide electric power to industrial units or residences

of the switch at the battery circuit and u_3 which is the duty cycle of the switch at the supercapacitor circuit.

Next, the state-space model of the DC microgrid is written in the following matrix form

$$
\begin{pmatrix} \dot{x}_1 \\ \dot{x}_2 \\ \dot{x}_3 \\ \dot{x}_4 \\ \dot{x}_5 \\ \dot{x}_6 \\ \dot{x}_7 \\ \dot{x}_8 \\ \dot{x}_9 \end{pmatrix}
=
\begin{pmatrix}
-\frac{1}{R_1 C_1} x_1 - \frac{1}{C_1} x_3 + \frac{1}{R_1 C_1} V_{PV} \\
-\frac{1}{R_2 C_2} x_2 - \frac{1}{C_2} x_3 + \frac{1}{R_2 C_2} x_9 \\
-\frac{1}{R_4 C_4} x_4 - \frac{1}{C_4} x_6 + \frac{1}{R_4 C_4} V_B \\
\frac{1}{L_3}(x_1 - x_2 - R_{o_1} x_3) + \frac{1}{L_3} x_2 \\
-\frac{1}{R_5 C_5} x_5 + \frac{1}{C_5} x_6 + \frac{1}{R_5 C_5} x_9 \\
\frac{1}{L_6} x_4 - \frac{1}{L_6} x_5 - \frac{R_{o4}}{L_6} \\
-\frac{1}{R_7 C_7} x_7 + \frac{1}{C_7} x_8 + \frac{1}{R_7 C_7} x_9 \\
-\frac{R_{08}}{L_8} x_8 - \frac{1}{L_8} x_7 \\
\left(\frac{x_2 - x_9}{R_2 C_9} + \frac{x_5 - x_9}{R_5 C_9} + \frac{x_7 - x_9}{R_7 C_9} - \frac{x_9}{R_L C_9} \right)
\end{pmatrix}
+
\begin{pmatrix}
0 & 0 & 0 \\
-\frac{1}{C_2} x_3 & 0 & 0 \\
\frac{(x_2 + (R_{o_1} - R_{o_2}) x_3)}{L_3} & 0 & 0 \\
0 & 0 & 0 \\
0 & -\frac{1}{C_5} x_6 & 0 \\
0 & \frac{1}{L_6} x_5 & 0 \\
0 & 0 & 0 \\
0 & 0 & \frac{1}{L_8} V_s \\
0 & 0 & 0
\end{pmatrix}
\begin{pmatrix} u_1 \\ u_2 \\ u_3 \end{pmatrix}
$$

(3.10)

Consequently, the state-space model of DC microgrid is written in the following affine-in-the-input state-space form

$$\dot{x} = f(x) + g(x)u \tag{3.11}$$

where $x \in R^{9 \times 1}$, $u \in R^{3 \times 1}$, $f(x) \in R^{9 \times 1}$ and $g(x) \in R^{9 \times 3}$.

3.1.2.2 Faults and cyberattacks affecting the power system

Indicative faults and cyberattacks or malignant human intrusions that may affect the DC microgrid, which comprises the photovoltaic power unit, batteries and a

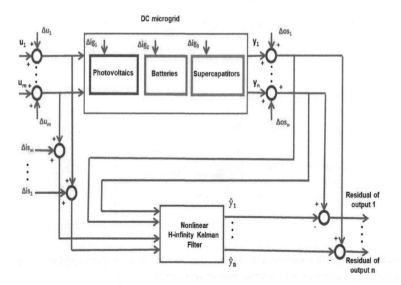

Figure 3.2   Faults and cyberattacks affecting the DC microgrid that comprises a photovoltaic power unit, batteries and a supercapacitor

supercapacitor, are shown in Fig. 3.2. The Kalman Filter's outputs are compared at each sampling instance against the outputs measured from the DC microgid thus providing the residuals' sequence. These faults are: (i) faults at the control input gains $\Delta i g_i$, $i = 1. \cdots, 3$ associated with the photovoltaic power unit, the batteries and the supercapacitor, (ii) faults (additive disturbances) at the control inputs Δu_i, $i = 1, \cdots, 3$ of the DC microgrid, (iii) faults at the control input sensors $\Delta i s_i$, $i = 1, \cdots, 3$ of the DC microgrid, (iv) faults at the output measurement sensors $\Delta o s_i$, $i = 1, \cdots, 9$ of the microgrid and (v) cyber-attacks to the control and data acquisition software of the DC microgrid.

3.1.3   APPROXIMATE LINEARIZATION OF THE STATE-SPACE MODEL OF THE DC MICROGRID

The dynamic model of the DC microgrid undergoes approximate linearization around the temporary operating point (x^*, u^*), which is recomputed at each sampling period. This temporary operating point is defined by x^*, which is the present value of the state vector of the DC microgrid and by u^* which is the last sampled value of the control inputs vector. The lineariation procedure makes use of Taylor series expansion and requires the computation of the associated Jacobian matrices. The linearization procedure gives the new state-space model:

$$\dot{x} = Ax + Bu + \tilde{d} \tag{3.12}$$

where $\tilde{d}$ is the cumulative disturbance term, while it holds that

$$A = \nabla_x[f(x) + g_1(x)u_1 + g_2(x)u_2 + g_3(x)u_3]\,|_{(x^*,u^*)} \Rightarrow$$
$$A = \nabla_x[f(x)]\,|_{(x^*,u^*)} + \nabla_x[g_1(x)]u_1\,|_{(x^*,u^*)} + \tag{3.13}$$
$$+ \nabla_x[g_2(x)]u_2\,|_{(x^*,u^*)} + \nabla_x[g_3(x)]u_3\,|_{(x^*,u^*)}$$

$$B = \nabla_u[f(x) + g_1(x)u_1 + g_2(x)u_2 + g_3(x)u_3]\,|_{(x^*,u^*)} \Rightarrow B = g(x)\,|_{(x^*,u^*)} \tag{3.14}$$

The computation of the Jacobian matrix $\nabla_x[f(x)]\,|_{(x^*,u^*)}$ gives:

$$\nabla_x[f(x)]\,|_{(x^*,u^*)}=$$

$$=\begin{pmatrix}
-\frac{1}{R_1C_1} & 0 & -\frac{1}{C_3} & 0 & 0 & 0 & 0 & 0 & 0 \\
0 & -\frac{1}{R_2C_2} & \frac{1}{C_3} & 0 & 0 & 0 & 0 & 0 & 0 \\
\frac{1}{L_3} & -\frac{1}{L_3} & -\frac{R_{o1}}{L_3} & 0 & 0 & 0 & 0 & 0 & 0 \\
0 & 0 & 0 & -\frac{1}{R_4C_4} & 0 & -\frac{1}{C_4} & 0 & 0 & 0 \\
0 & 0 & 0 & 0 & -\frac{1}{R_5C_5} & -\frac{1}{C_5} & 0 & 0 & \frac{1}{R_5C_5} \\
0 & 0 & 0 & \frac{1}{L_6} & -\frac{1}{L_6} & -\frac{R_{o4}}{L_6} & 0 & 0 & 0 \\
0 & 0 & 0 & 0 & 0 & 0 & -\frac{1}{R_7C_7} & \frac{1}{C_7} & \frac{1}{R_7C_7} \\
0 & 0 & 0 & 0 & 0 & 0 & -\frac{1}{L_8} & -\frac{R_{o8}}{L_8} & 0 \\
0 & \frac{1}{C_9R_2} & 0 & 0 & \frac{1}{C_9R_5} & 0 & \frac{1}{C_9R_7} & 0 & \frac{\partial f_9}{\partial x_9}
\end{pmatrix}$$
$$\tag{3.15}$$

where $\frac{\partial f_9}{\partial x_9} = -\frac{1}{C_9R_2} - \frac{1}{C_9R_5} - \frac{1}{C_9R_7} - \frac{1}{R_L}$

Furthermore, using that the control input gains matrix $g(x) \in R^{9\times3}$ is also written as $g(x) = [g_1(x)\ g_2(x)\ g_3(x)]$ where $g_1(x) \in R^{9\times1}$, $g_2(x) \in R^{9\times1}$ and $g_3(x) \in R^{9\times1}$, one proceeds to the computation of the Jacobian $\nabla_x[g_1(x)]\,|_{(x^*,u^*)}$

$$\nabla_x[g_1(x)]\,|_{(x^*,u^*)}=\begin{pmatrix}
0 & 0 & 0 & 0 & 0 & 0 & 0 & 0 & 0 \\
0 & 0 & -\frac{1}{C_2} & 0 & 0 & 0 & 0 & 0 & 0 \\
0 & \frac{1}{L_3} & \frac{R_{o1}-R_{o2}}{L_3} & 0 & 0 & 0 & 0 & 0 & 0 \\
0 & 0 & 0 & 0 & 0 & 0 & 0 & 0 & 0 \\
0 & 0 & 0 & 0 & 0 & 0 & 0 & 0 & 0 \\
0 & 0 & 0 & 0 & 0 & 0 & 0 & 0 & 0 \\
0 & 0 & 0 & 0 & 0 & 0 & 0 & 0 & 0 \\
0 & 0 & 0 & 0 & 0 & 0 & 0 & 0 & 0 \\
0 & 0 & 0 & 0 & 0 & 0 & 0 & 0 & 0
\end{pmatrix} \tag{3.16}$$

as well as to the computation of the Jacobian matrix $\nabla_x[g_2(x)] \mid_{(x^*,u^*)}$

$$\nabla_x g_2[x)] \mid_{(x^*,u^*)} = \begin{pmatrix} 0 & 0 & 0 & 0 & 0 & 0 & 0 & 0 & 0 \\ 0 & 0 & 0 & 0 & 0 & 0 & 0 & 0 & 0 \\ 0 & 0 & 0 & 0 & 0 & 0 & 0 & 0 & 0 \\ 0 & 0 & 0 & 0 & 0 & 0 & 0 & 0 & 0 \\ 0 & 0 & 0 & 0 & 0 & -\frac{1}{C_5} & 0 & 0 & 0 \\ 0 & 0 & 0 & 0 & \frac{1}{L_6} & 0 & 0 & 0 & 0 \\ 0 & 0 & 0 & 0 & 0 & 0 & 0 & 0 & 0 \\ 0 & 0 & 0 & 0 & 0 & 0 & 0 & 0 & 0 \\ 0 & 0 & 0 & 0 & 0 & 0 & 0 & 0 & 0 \end{pmatrix} \tag{3.17}$$

while about the computation of the Jacobian matrix $\nabla_x[g_3(x)] \mid_{(x^*,u^*)}$ it holds:
$\nabla_x g_2[x)] \mid_{(x^*,u^*)} = 0_{9 \times 9}$.

For the DC microgrid a stabilizing (H-infinity) feedback controller is

$$u(t) = -Kx(t) \tag{3.18}$$

with $K = \frac{1}{r}B^T P$, where P is a positive definite symmetric matrix, which is obtained from the solution of the Riccati equation

$$A^T P + PA + Q - P(\frac{2}{r}BB^T - \frac{1}{\rho^2}LL^T)P = 0 \tag{3.19}$$

where Q is a positive semi-definite symmetric matrix.

3.1.4 STATISTICAL FAULT DIAGNOSIS WITH THE H-INFINITY KALMAN FILTER

3.1.4.1 The H-infinity Kalman Filter

The H-infinity Kalman Filter is taken to represent the fault-free functioning of the DC microgrid. By comparing the filter's output against the real outputs of the DC microgrid, the residuals sequence is generated, which in turn can be used for fault diagnosis purposes. Filters designed to minimize a weighted norm of state errors are called H_∞ or minimax filters [198], [195], [78], [226]. The discrete time H_∞ filter uses the same state-space model as the Kalman Filter, which has the form

$$x(k+1) = A(k)x(k) + B(k)u(k) + w(k)$$
$$z(k) = C(k)x(k) + v(k) \tag{3.20}$$

where $x \in R^{n \times 1}$, $A \in R^{n \times n}$, $B \in R^{n \times m}$, $u \in R^{m \times 1}$, $w \in R^{n \times 1}$, $z \in R^{q \times 1}$, $C \in R^{q \times n}$ and $v \in R^{q \times 1}$. Moreover, $E[w(k)] = 0$, $E[w(k)w(k)^T] = Q(k) = diag[\delta ii]$, $E[v(k)] = 0$, $E[v(k)v(k)^T] = R(k) = diag[\delta_{ii}]$ and $E(w(k)v(k)^T) = 0$. The update of the state estimate is again given by

$$\hat{x}(k) = \hat{x}^-(k) + K(k)(z(k) - C(k)\hat{x}^-(k)) \tag{3.21}$$

This estimation process minimizes the trace of the covariance matrix of the state vector estimation error

$$J = \frac{1}{2}E\{\tilde{x}(k)^T \cdot \tilde{x}(k)\} = \frac{1}{2}tr(P^-(k)) \tag{3.22}$$

where $\tilde{x}^-(k) = x(k) - \hat{x}^-(k)$ and $P^-(k) = E[\tilde{x}^-(k)^T \cdot \tilde{x}^-(k)]$. The H_∞ filtering approach defines first a transformation $d(k) = L(k)x(k)$, where $L(k) \in R^{n \times n}$ is a full rank matrix. The use of the aforementioned transformation allows certain combinations of states to be given more weight than others [198], [195], [78-226]. By denoting as $A(k)$, $B(k)$ and $C(k)$ the discrete-time equivalents of matrices A, B and C of the linearized state-space model of the system, the recursion of the H-infinity Kalman Filter can be formulated again in terms of a *measurement update* and a *time update* part [78-226]

Measurement update:

$$D(k) = [I - \theta W(k)P^-(k) + C^T(k)R(k)^{-1}C(k)P^-(k)]^{-1}$$
$$K(k) = P^-(k)D(k)C^T(k)R(k)^{-1} \tag{3.23}$$
$$\hat{x}(k) = \hat{x}^-(k) + K(k)[y(k) - C\hat{x}^-(k)]$$

Time update:

$$\hat{x}^-(k+1) = A(k)x(k) + B(k)u(k)$$
$$P^-(k+1) = A(k)P^-(k)D(k)A^T(k) + Q(k) \tag{3.24}$$

where it is assumed that the parameter θ is sufficiently small to maintain

$$P^-(k)^{-1} - \theta W(k) + C^T(k)R(k)^{-1}C(k) \tag{3.25}$$

positive definite. When $\theta = 0$ the H_∞ Kalman Filter becomes equivalent to the standard Kalman Filter. It is noted that apart from the process noise covariance matrix $Q(k)$ and the measurement noise covariance matrix $R(k)$ the H_∞ Kalman filter requires tuning of the weight matrices L and S, as well as of parameter θ.

To elaborate on the matrices which appear in the *Measurement update* part and in the *Time update part* of the H-infinity Kalman Filter, the following can be noted: Matrix $R(k)$ is the measurement noise covariance matrix, that is the covariance matrix of the measurement error vector of the system. Matrix $P^-(k)$ is the a-priori state vector estimation error covariance matrix of the system, that is the covariance matrix of the state vector estimation error prior to receiving the updated measurement of the system's outputs. Matrix $W(k)$ is a weight matrix which defines the significance to be attributed by the H-infinity Kalman Filter to the minimization the state vector's estimation error, relatively to the effects that the noise affecting the system may have.

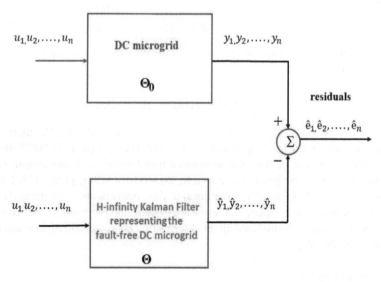

Figure 3.3 Residuals'generation for the DC microgrid, with the use of Kalman Filtering

Finally, matrix $D(k)$ stands for a modified a-posteriori state vector estimation error covariance matrix, that is the covariance matrix of the state vector estimation error after receiving the updated measurement of the system's outputs. Conclusively, the H-infinity Kalman Filter retains the structure of the typical Kalman Filter, that is a recursion in discrete-time comprising a Time update part (computation of variables prior to receiving measurements) and a Measurement update part (computation of variables after measurements have been received). There is a modified a-posteriori state vector estimation error covariance matrix, which in turn takes into account a weight matrix that defines the accuracy of the state estimation under the effects of elevated noise.

3.1.4.2   Fault detection

The residuals' sequence, that is the differences between (i) the real outputs of the DC microgrid and (ii) the outputs estimated by the Kalman Filter (Fig. 3.3) is a discrete error process e_k with dimension $m \times 1$ (here $m = N$ is the dimension of the output measurements vector). Actually, it is a zero-mean Gaussian white-noise process with covariance given by E_k.

A conclusion can be stated based on a measure of certainty that the DC microgrid has neither been subjected to a fault nor to a cyberattack. To this end, the following *normalized error square* (NES) is defined [195]

$$\varepsilon_k = e_k^T E_k^{-1} e_k \qquad (3.26)$$

The normalized error square follows a χ^2 distribution. An appropriate test for the normalized error sum is to numerically show that the following condition is met within a level of confidence (according to the properties of the χ^2 distribution)

$$E\{\varepsilon_k\} = m \tag{3.27}$$

This can be achieved using statistical hypothesis testing, which is associated with confidence intervals. For instance, a 95% confidence interval is frequently applied, which is specified using $100(1 - a)$ with $a = 0.05$. Actually, a two-sided probability region is considered cutting-off two end tails of 2.5% each. For M runs the normalized error square, that is obtained from the residuals' sequence, is given by

$$\bar{\varepsilon}_k = \frac{1}{M}\sum_{i=1}^{M}\varepsilon_k(i) = \frac{1}{M}\sum_{i=1}^{M}e_k^T(i)E_k^{-1}(i)e_k(i) \tag{3.28}$$

where ε_i stands for the i-th run at time t_k. Then $M\bar{\varepsilon}_k$ will follow a χ^2 density with Mm degrees of freedom. This condition can be checked using a χ^2 test. The hypothesis holds, if the following condition is satisfied

$$\bar{\varepsilon}_k \in [\zeta_1, \zeta_2] \tag{3.29}$$

where ζ_1 and ζ_2 are derived from the tail probabilities of the χ^2 density.

3.1.4.3 Fault isolation

By applying the statistical test into the individual components of the DC microgrid, it is also possible to find out the specific component that has been subjected to a fault or cyber-attack [198], [195]. For a DC microgrid of n parameters suspected for change, one has to carry out n χ^2 statistical change detection tests, where each test is applied to the subset that comprises parameters $i - 1$, i and $i + 1$, $i = 1, 2, \cdots, n$. Actually, out of the n χ^2 statistical change detection tests, the ones that exhibit the highest score are those that identify the parameter that has been subjected to change.

In the case of multiple faults one can identify the subset of parameters that has been subjected to change by applying the χ^2 statistical change detection test according to a combinatorial sequence. This means that

$$\binom{n}{k} = \frac{n!}{k!(n-k)!} \tag{3.30}$$

tests have to take place, for all clusters in the DC microgrid, that finally comprise $n, n - 1, n - 2, \cdots, 2, 1$ parameters. Again the χ^2 tests that give the highest scores indicate the parameters which are most likely to have been subjected to change.

3.1.5 SIMULATION TESTS

To perform fault detection and isolation for the DC microgrid the statistical properties of the residuals of the H-infinity Kalman Filter have been used. The nominal parameters of the model of the DC microgrid which have been used to simulate its functioning in the fault-free condition are given in [98], [100]. The simulated faults were associated with parametric changes up to 10% of the nominal values of the parameters of the DC microgrid. The dimension of the measurements vector was equal to $n = 9$, so was also the dimension of the residuals vector. This signifies that in the fault-free case the considered statistical test should have a value that is very close to the mean value of the χ^2 distribution with $n = 9$ degrees of freedom. Consequently, in the fault-free functioning of the DC microgrid the statistical test should have a value equal to 9. The 98% confidence intervals about finding the system in the fault-free case were determined by the lower and upper bounds $L = 8.85$ and $U = 9.15$ respectively. As confirmed by the simulation experiments, as long as the value of the statistical test falls within these confidence intervals, it is concluded that the functioning of the DC microgrid is normal. On the other side, when the above noted upper or lower bounds are exceeded it can be concluded that the system has been subjected to a failure and an alarm can be launched. The no-fault case is depicted in Fig. 3.4. Furthermore, in Fig. 3.5 to Fig. 3.7 the performance of the statistical test is shown in the case that a fault (cyber-attack) affects the control inputs (duty-cycles) u_i $i = 1, 2, 3$ of the microgrid. Additionally, in Fig. 3.8 to Fig. 3.10 the performance of the statistical test is shown in case that a fault affects the voltage sources of the three DC circuits that constitute the microgrid.

The H-infinity Kalman Filter is expected to have improved estimation accuracy comparing to the Extended Kalman Filter, when the estimation process takes place under elevated measurement noise levels. On the other side, it is important that the proposed fault threshold definition approach is generic and can be applied to all state-observers or filters which may be used for representing the fault-free functioning of the monitored DC-microgrid and for generating the associated sequence of residuals. Actually, it holds that the sum of the squares of the residuals being weighted by the associated inverse covariance matrix is a stochastic variable which follows the χ^2 distribution with degrees of freedom being equal to the dimension of the residuals vector. In the fault-free case this stochastic variable will coincide with the mean value of the above-noted χ^2 distribution. In general, taking also into account the randomness of the measurement noise, this stochastic variable should be found within the confidence intervals of the χ^2-square distribution. For instance, by selecting fault thresholds associated with the 96% or the 98% confidence intervals of the χ^2 distribution, one can arrive at a highly reliable and almost infallible fault diagnosis test. Consequently, by using the confidence intervals of the χ^2 distribution one can define precise fault thresholds which demonstrate in a reliable manner the occurrence of faults and parametric changes.

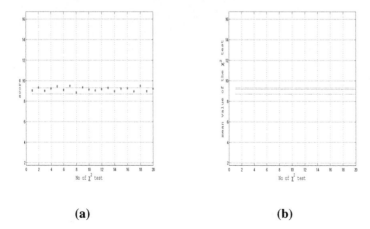

(a)   (b)

Figure 3.4 (a) Values of successive χ^2 tests when no fault exists at the DC microgrid and (b) mean value of the χ^2 tests when no fault exists at the DC microgrid and (the horizontal axes depicts the sequence of the performed test and the vertical axes show the values of test variable)

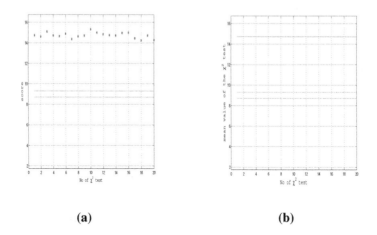

(a)   (b)

Figure 3.5 (a) Values of successive χ^2 tests when a fault affects control input (duty cycle) u_1 and (b) mean value of the χ^2 tests when a fault affects control input (duty cycle) u_1 (the horizontal axes depicts the sequence of the performed test and the vertical axes show the values of test variable)

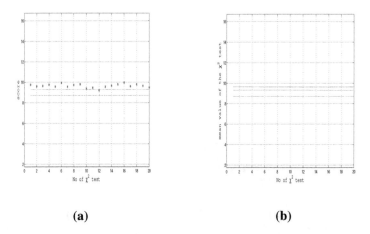

(a)   **(b)**

Figure 3.6   (a) Values of successive χ^2 tests when a fault affects control input (duty cycle) u_2 and (b) mean value of the χ^2 tests when a fault affects control input (duty cycle) u_2 (the horizontal axes depict the sequence of the performed test and the vertical axes show the values of test variable)

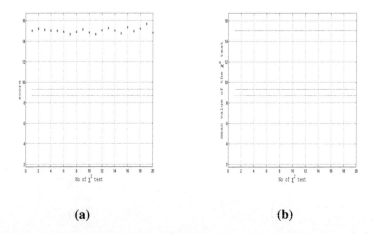

(a)   **(b)**

Figure 3.7   (a) Values of successive χ^2 tests when a fault affects control input (duty cycle) u_3 and (b) mean value of the χ^2 tests when a fault affects control input (duty cycle) u_3 (the horizontal axes depict the sequence of the performed test and the vertical axes show the values of test variable)

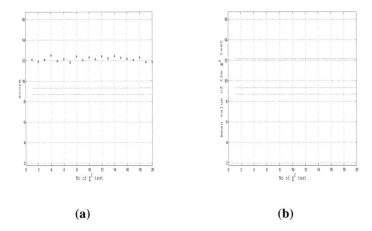

(a) (b)

Figure 3.8 (a) Values of successive χ^2 tests when a fault affects the voltage source V_{PV} and (b) mean value of the χ^2 tests when a fault affects the voltage source V_{PV} (the horizontal axes depict the sequence of the performed test and the vertical axes show the values of test variable)

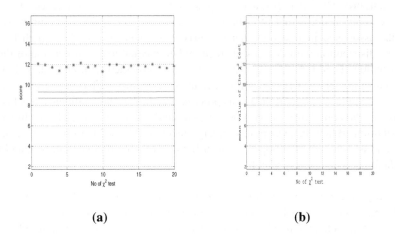

(a) (b)

Figure 3.9 (a) Values of successive χ^2 tests when a fault affects the voltage source V_b and (b) mean value of the χ^2 tests when a fault affects the voltage source V_b (the horizontal axes depict the sequence of the performed test and the vertical axes show the values of test variable)

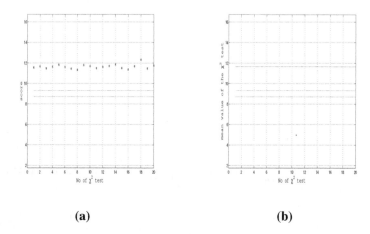

(a)   (b)

Figure 3.10   (a) Values of successive χ^2 tests when a fault affects the voltage source V_{sc} and (b) mean value of the χ^2 tests when a fault affects the voltage source V_{sc} (the horizontal axes depict the sequence of the performed test and the vertical axes show the values of test variable)

The proposed fault threshold definition is generic and can be applied to all state-observers or filters which may be used for representing the fault-free functioning of the monitored DC-microgrid and for generating the associated sequence of residuals. Actually, it holds that the sum of the squares of the residuals being weighted by the associated inverse covariance matrix is a stochastic variable which follows the χ^2 distribution with degrees of freedom being equal to the dimension of the χ^2 distribution. Therefore this stochastic variable, is bounded without undergoing any normalization. Narrow (wide) confidence intervals signify high (low) certainty about the correctness of the fault diagnosis process. The higher the confidence level is, the narrower the confidence interval becomes around the mean value of the chi-square distribution. When the system has undergone a fault, the value of the statistical test exceeds significantly, the upper or lower bound of the confidence interval, and thus the fault becomes clearly visible.

The proposed statistical condition monitoring method consists of two parts: (i) the fault detection test and (ii) the fault isolation test. In fault detection, the statistical test uses the complete residuals' vector and demonstrates whether the monitored system in its entirety remains in the fault-free condition or if it has undergone a parametric change. In fault isolation, the statistical test is applied in subspaces defined by parts of the residuals' vector. A condition that should hold is that the fault is not propagated throughout the entire system and thus the changed parameter affects only specific elements of the residuals' vector. Then, by performing the statistical test in all possible subspaces and by sorting the tests that give the highest score one can isolate the source of the fault. There is also the case in which specific parametric changes are related with singularities of the statistical test's covariance matrix and consequently the test cannot be performed and the associated faults become non-identifiable.

3.2 FAULT DIAGNOSIS FOR ELECTRICALLY ACTUATED GAS COMPRESSORS

3.2.1 OUTLINE

At a second stage, the chapter treats the problem of fault diagnosis for electrically actuated gas compressors. Due to harsh and variable operating conditions, natural gas processing units often undergo failures and the solution of the associated fault detection and isolation (FDI) problems is a meaningful and nontrivial task [159], [140], [80], [231]. As noted before, centrifugal gas-compressors are widely used in the natural gas processing industry, as for instance in gas terminals, in the loading or unloading of LNG ships. and in the supply of gas distribution networks with natural gas at the specified pressure levels [86], [142], [241], [34]. Such centrifugal gas compressors are mechanically actuated by gas turbines or by electric motors. In case of electric actuation, synchronous or asynchronous (induction) three-phase motors are commonly used [17], [240], [83]. More recently, the use of multi-phase motors for the actuation of gas compressors has been also considered [288], [185], [244]. Multiphase motors have been employed for long in applications where high power and high torque is needed [64], [65], [148]. By distributing the required power in a large number of phases, the power load of each individual phase is reduced [212], [27], [129]. Consequently, the associated power electronics (voltage source converters) function also at smaller voltages and currents. Another feature is that the cumulative rates of power in multiphase machines can be raised without stressing the connected converters [168], [169], [116]. Furthermore, the frequency of PWM inputs can be increased while the amplitude of such inputs can be reduced, and this signifies avoidance of mechanical vibrations during the functioning of the motor. Multiphase motors are also fault tolerant because such machines remain functional even if failures affect certain phases [208], [8].

Five-phase PM synchronous motors and five-phase asynchronous induction motors are among the types of multiphase motors one can consider for the actuation of gas compressors [261]. For ensuring the reliable performance of such multiphase motors, elaborated nonlinear control methods have been proposed [11] ,[9], [10], [257], [66]. Moreover, fault diagnosis methods for gas-compressors have been introduced [159], [140], [80], [231]. In the present chapter, a novel fault diagnosis technique has been developed for the integrated gas compression system, which is electrically actuated by a five-phase induction motor. The dynamic model of the gas-compressor and five-phase IM undergoes first approximate linearization around the temporary operating point (x^*, u^*) using first-order Taylor series expansion and the computation of the associated Jacobian matrices [188], [18], [189]. The linearization process takes place at each sampling instant and the linearization point is defined by the present value of the system's state vector x^* and by the last sampled value of the control inputs vector u^* [206], [195], [198], [204], [191], [243]. Next, to perform state estimation for the gas-compression system, the H-infinity Kalman Filter is used as a robust state estimator.

By applying the H-infinity Kalman Filter, the chapter treats also the fault diagnosis problem for the integrated system that comprises the centrifugal gas-compressor, which is actuated by the five-phase induction motor [206], [195], [198], [204]. To this end, statistical processing of the residuals of the aforementioned H-infinity Kalman Filter has been performed. Actually, a sequence of residual vectors has been generated by measuring at each sampling instance the differences between the real values of the outputs of the gas-compression system and the estimated outputs which have been provided by the H-infinity Kalman Filter. It has been used that the sum of the squares of the residuals' vectors, being weighted by the inverse of the associated covariance matrix, is a stochastic variable (statistical test) which follows the χ^2 distribution with degrees of freedom and mean value which are equal to the dimension of the residuals' vector. Furthermore, the confidence intervals of this χ^2 distribution have been used to define fault thresholds which allow for deciding with a high level of confidence (95% or more) about the appearance of a fault in the gas compression system. Whenever the previously noted statistical test exceeds these confidence intervals, the existence of a fault can be concluded. Moreover, by applying the statistical test in subspaces of the state-space model of the gas-compression system fault isolation can be also performed.

3.2.2 DYNAMIC MODEL OF THE GAS-COMPRESSOR AND FIVE-PHASE INDUCTION MOTOR SYSTEM

The diagram of the gas-compressor which is actuated by a five-phase induction motor is shown again in Fig. 3.11. Through the inlet valve which is denoted by gain K_i, the gas is fed into the inlet tank which is described by pressure variable P_i. Next, the gas passes through the compressor at a mass flow rate which is denoted as m. The rotational motion of the compressor is due to the torque, which is provided by a five-phase induction motor. The angular speed of the compressor and the five-phase induction motor is defined as ω. The gas that comes out of the compressor is stored in the outlet tank at a pressure which is denoted as P_o. In the output of the outlet tank there is an outlet valve which is described by gain K_o. Moreover, it is possible to recycle part of the gas of the outlet tank back to the inlet tank, through a valve that is denoted by gain K_r.

The state vector of the compressor is defined once again as

$$x = [x_1, x_2, x_3, x_4, x_5]^T = [P_i, P_o, m, \omega, m_r]^T \tag{3.31}$$

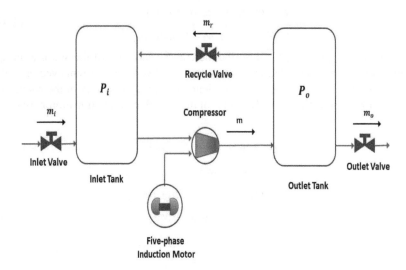

Figure 3.11   Diagram of the integrated gas compression system which comprises a centrifugal gas-compressor actuated by a five-phase induction motor.

while the state equations of the centrifugal compressor are given by

$$\dot{x}_1 = \frac{a_i^2}{V_i}[K_i\sqrt{p_{atm} - x_1} - x_3 + x_5]$$
$$\dot{x}_2 = \frac{a_o^2}{V_o}[-K_o\sqrt{x_2 - p_{atm}} + x_3 - x_5]$$
$$\dot{x}_3 = \frac{1}{L}[\pi(x_3, x_4)x_1 - x_2] \qquad (3.32)$$
$$\dot{x}_4 = \frac{1}{J}[\tau_d - \tau_c(x_3, x_4) - vx_4]$$
$$\dot{x}_5 = \frac{1}{T_r}[K_r\sqrt{x_2 - x_1} - K_i\sqrt{p_{atm} - x_1}]$$

The atmospheric pressure is denoted as p_{atm}. The compressor's characteristic function is denoted as $\pi(x_3, x_4)$ and is taken to be a polynomial function of state variables x_3 and x_4.

Equivalently, the compressor's torque is denoted as $\tau_c(x_3, x_4)$ and is also taken to be a polynomial function of state variables x_3 and x_4. The torque which enables the rotational motion of the compressor is denoted as τ_d and coincides with the electromagnetic torque of the five-phase induction motor. The five-phase induction motor, which is fed by a Voltage Source Inverter is shown in the diagram of Fig. 3.12.

The state vector of the five-phase induction motor is given by

$$x = [\omega_r, x_6, x_7, x_8, x_9, x_{10}, x_{11}] \Rightarrow$$
$$x = [\omega_r, i_{s,a}, i_{s,\beta}, \psi_{r,a}, \psi_{r,\beta}, i_{s,x}, i_{s,y}]^T \qquad (3.33)$$

In this notation $i_{s,a}, i_{s,\beta}, i_{s,x}, i_{s,y}$ describe the current variables of the stator after applying a generalized Clarke's transformation to the five-phase currents vector of the motor. Moreover, $\psi_{r,a}, \psi_{r,\beta}$ are the coefficients of the magnetic flux of the rotor. The state equations that describe the dynamics of the five-phase induction motor are given next

$$\dot{\omega}_r = m_1 T_e - m_1 T_L - m_2 \omega_r$$
$$\dot{x}_6 = aA_r x_8 + a\omega_r x_9 - bx_6 + cv_{s,a}$$
$$\dot{x}_7 = aA_r x_9 - a\omega_r x_8 - bx_7 + cv_{s,b}$$
$$\dot{x}_8 = -A_r x_8 - \omega_r x_9 + MA_r x_6 \qquad (3.34)$$
$$\dot{x}_9 = -A_r x_9 + \omega_r x_8 + MA_r x_7$$
$$\dot{x}_{10} = dx_{10} + ev_{s,x}$$
$$\dot{x}_{11} = dx_{11} + ev_{s,y}$$

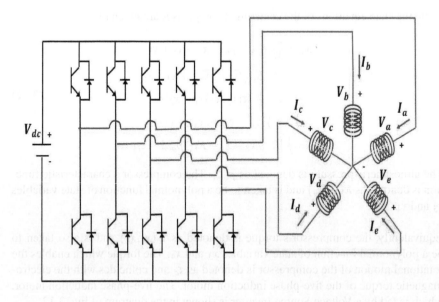

Figure 3.12 A VSI-fed five-phase induction motor giving actuation to the gas-compressor

The electromagnetic torque of the five-phase motor is given by:

$$T_e = \frac{PM}{L_r}(\psi_{r,a}i_{s,b} - \psi_{r,b}i_{s,a}) \Rightarrow T_e = \frac{PM}{L_r}(x_8 x_7 - x_9 x_6) \tag{3.35}$$

The coefficients of the dynamic model of the five-phase induction motor are: $a = \frac{M}{\sigma L_s L_r}$, $A_r = \frac{1}{\tau_r}$, $\tau_r = \frac{L_r}{R_r}$, $b = \frac{R_s}{\sigma L_s} + \frac{R_r M^2}{\sigma L_s L_r^2}$, $\sigma = 1 - \frac{M^2}{L_s L_r}$, $c = \frac{1}{\sigma L_s}$, $d = -\frac{R_s}{L_{1s}}$, $e = \frac{1}{L_{1s}}$, $m_1 = \frac{P}{J}$, $m_2 = \frac{B_m}{J}$.

In the above coefficients R_r and L_r are the resistance and the inductance of the rotor, R_s and L_s are the resistance and the inductance of the stator, M is the mutual inductance between the stator and the rotor and L_{1s} is the leakage induction of the stator.

In the following, the integrated dynamic model of the electrically actuated gas compression system will comprise both the dynamics of the gas-compressor and the dynamics of the five-phase induction motor. It is also possible to further extend this model by including in it the dynamics of the five-phase voltage source inverter which generates the control inputs of the induction motor (Fig. 3.11).

By connecting the gas-compressor with the five-phase induction motor, the turn speed of the motor coincides with the one of the compressor, that is $\omega = \omega_r$. Besides, the torque τ_d which activates the compressor coincides with the electromagnetic torque of the motor, that is $\tau_d = T_e$. Thus, the integrated model of the gas compression system becomes:

$$\dot{x}_1 = \frac{a_i^2}{V_i}[K_i\sqrt{p_{atm} - x_1} - x_3 + x_5]$$
$$\dot{x}_2 = \frac{a_o^2}{V_o}[-K_o\sqrt{x_2 - p_{atm}} + x_3 - x_5]$$
$$\dot{x}_3 = \frac{1}{L}[\pi(x_3, x_4)x_1 - x_2]$$
$$\dot{x}_4 = \frac{1}{J}[\tau_d - \tau_c(x_3, x_4) - vx_4]$$
$$\dot{x}_5 = \frac{1}{T_r}[K_r\sqrt{x_2 - x_1} - K_i\sqrt{p_{atm} - x_1}]$$
$$\dot{x}_6 = aA_r x_8 + a\omega_r x_9 - bx_6 + cv_{s,a} \tag{3.36}$$
$$\dot{x}_7 = aA_r x_9 - a\omega_r x_8 - bx_7 + cv_{s,b}$$
$$\dot{x}_8 = -A_r x_8 - \omega_r x_9 + MA_r x_6$$
$$\dot{x}_9 = -A_r x_9 + \omega_r x_8 + MA_r x_7$$
$$\dot{x}_{10} = dx_{10} + ev_{s,x}$$
$$\dot{x}_{11} = dx_{11} + ev_{s,y}$$

Moreover, using the following notation about the aggregate control inputs vector of the system

$$u = [u_1, u_2, u_3, u_4, u_5, u_6]^T = [K_i, K_r, v_{s,a}, v_{s,b}, v_{s,x}, v_{s,y}]^T \tag{3.37}$$

the state-space model of the integrated gas-compressor and five-phase induction motor can be also given in the form

$$
\begin{pmatrix} \dot{x}_1 \\ \dot{x}_1 \\ \dot{x}_3 \\ \dot{x}_4 \\ \dot{x}_5 \\ \dot{x}_6 \\ \dot{x}_7 \\ \dot{x}_8 \\ \dot{x}_9 \\ \dot{x}_{10} \\ \dot{x}_{11} \end{pmatrix} = \begin{pmatrix} \frac{a_i^2}{V_i}(-x_3 + x_5) \\ \frac{a_o^2}{V_o}\left[-K_o\sqrt{x_2 - p_{atm}} + x_3 - x_5\right] \\ \frac{1}{L}[\pi(x_3,x_4)x_1 - x_2] \\ \frac{1}{J}[\tau_d - \tau_c(x_3,x_4) - vx_4] \\ 0 \\ aA_r x_8 + a\omega_r x_9 - b x_6 \\ aA_r x_9 - a\omega_r x_8 - b x_7 \\ -A_r x_8 - \omega_r x_9 + MA_r x_6 \\ -A_r x_9 + \omega_r x_8 + MA_r x_7 \\ d x_{10} \\ d x_{11} \end{pmatrix} +
$$

(3.38)

$$
+ \begin{pmatrix} \frac{a_i^2}{V_i}\sqrt{p_{atm} - x_1} & 0 & 0 & 0 & 0 & 0 \\ 0 & 0 & 0 & 0 & 0 & 0 \\ 0 & 0 & 0 & 0 & 0 & 0 \\ 0 & 0 & 0 & 0 & 0 & 0 \\ -\frac{1}{T_r}\sqrt{p_{atm} - x_1} & \frac{1}{T_r}\sqrt{x_2 - x_1} & 0 & 0 & 0 & 0 \\ 0 & 0 & c & 0 & 0 & 0 \\ 0 & 0 & 0 & c & 0 & 0 \\ 0 & 0 & 0 & 0 & 0 & 0 \\ 0 & 0 & 0 & 0 & 0 & 0 \\ 0 & 0 & 0 & 0 & e & 0 \\ 0 & 0 & 0 & 0 & 0 & e \end{pmatrix} \begin{pmatrix} u_1 \\ u_2 \\ u_3 \\ u_4 \\ u_5 \\ u_6 \end{pmatrix}
$$

The integrated system can be also written in the nonlinear affine-in-the-input state-space form

$$
\dot{x} = f(x) + g(x)u \tag{3.39}
$$

3.2.3   APPROXIMATE LINEARIZATION OF THE GAS-COMPRESSION SYSTEM

The dynamic model of the gas-compression system undergoes approximate linearization around the temporary operating point (x^*, u^*), where x^* is the present value of the system's state vector and u^* is the last sampled value of the control inputs vector. The linearization takes place at each sampling instant and is based on first-order Taylor series expansion. The modeling error which is due to truncation of higher-order terms from the Taylor series is viewed as a perturbation, which is asymptotically compensated by the robustness of the control algorithm.

The initial nonlinear model of the gas-compression system is in the nonlinear affine-in-the-input state-space form:

$$\dot{x} = f(x) + g(x)u \quad x \in R^{11 \times 1}, \; f(x) \in R^{11 \times 1}, \; g(x) \in R^{11 \times 6}, \; u \in R^{6 \times 1} \tag{3.40}$$

After linearization with the use of first order Taylor series expansion it is written in the equivalent linearized form

$$\dot{x} = Ax + Bu + \tilde{d} \tag{3.41}$$

where $\tilde{d}$ is the cumulative disturbances vector which may comprise (i) the modeling error due to truncation of higher-order terms from the Taylor series, (ii) exogenous perturbations and (iii) sensor measurement noise of any distribution. Matrices A and B are the Jacobian matrices of the system which are given by

$$A = \nabla_x[f(x) + g(x)u] \,|_{(x^*,u^*)} \Rightarrow A = \nabla_x[f(x)] \,|_{(x^*,u^*)} +$$

$$+ \nabla_x[g_1(x)u] \,|_{(x^*,u^*)} + \nabla_x[g_2(x)u] \,|_{(x^*,u^*)} +$$

$$+ \nabla_x[g_3(x)u] \,|_{(x^*,u^*)} + \nabla_x[g_4(x)u] \,|_{(x^*,u^*)} + \tag{3.42}$$

$$+ \nabla_x[g_5(x)u] \,|_{(x^*,u^*)} + \nabla_x[g_6(x)u] \,|_{(x^*,u^*)}$$

$$B = \nabla_u[f(x) + g(x)u] \,|_{(x^*,u^*)} \Rightarrow B = g(x) \,|_{(x^*,u^*)} \tag{3.43}$$

where $g_i(x)$, $i = 1, 2, \cdots, 6$ are the column vectors which constitute,

$$g(x) = [g_1(x), g_2(x), g_3(x), g_4(x), g_5(x), g_6(x)]^T \tag{3.44}$$

that is the system's control inputs gain matrix.

Next, the elements of the Jacobian matrices are computed. First, the computation of the Jacobian matrix $\nabla_x[f(x)] \,|_{(x^*,u^*)}$ is performed.

First row of the Jacobian matrix $\nabla_x[f(x)] \,|_{(x^*,u^*)}$: $\frac{\partial f_1}{\partial x_1} = 0$, $\frac{\partial f_1}{\partial x_2} = 0$, $\frac{\partial f_1}{\partial x_3} = -\frac{a_i^2}{V_i}$, $\frac{\partial f_1}{\partial x_4} = 0$, $\frac{\partial f_1}{\partial x_5} = \frac{a_i^2}{V_i}$, $\frac{\partial f_1}{\partial x_6} = 0$, $\frac{\partial f_1}{\partial x_7} = 0$, $\frac{\partial f_1}{\partial x_8} = 0$, $\frac{\partial f_1}{\partial x_9} = 0$, $\frac{\partial f_1}{\partial x_{10}} = 0$, $\frac{\partial f_1}{\partial x_{11}} = 0$.

Second row of the Jacobian matrix $\nabla_x[f(x)] \,|_{(x^*,u^*)}$: $\frac{\partial f_2}{\partial x_1} = 0$, $\frac{\partial f_2}{\partial x_2} = \frac{a_o*^2}{V_o}(-\frac{K_o}{2})(x_2 - P_{atm})^{-\frac{1}{2}}$, $\frac{\partial f_2}{\partial x_3} = \frac{a_o^2}{V_o}$, $\frac{\partial f_2}{\partial x_4} = 0$, $\frac{\partial f_2}{\partial x_5} = 0$, $\frac{\partial f_2}{\partial x_6} = -\frac{a_o^2}{V_o}$, $\frac{\partial f_2}{\partial x_7} = 0$, $\frac{\partial f_2}{\partial x_8} = 0$, $\frac{\partial f_2}{\partial x_9} = 0$, $\frac{\partial f_2}{\partial x_{10}} = 0$, $\frac{\partial f_2}{\partial x_{11}} = 0$.

Third row of the Jacobian matrix $\nabla_x[f(x)] \,|_{(x^*,u^*)}$: $\frac{\partial f_3}{\partial x_1} = \frac{1}{L}\pi(x_3, x_4)$, $\frac{\partial f_3}{\partial x_2} = -\frac{1}{L}$, $\frac{\partial f_3}{\partial x_3} = \frac{1}{L}\frac{\partial \pi(x_3, x_4)}{\partial x_3}x_1$, $\frac{\partial f_3}{\partial x_4} = \frac{1}{L}\frac{\partial \pi(x_3, x_4)}{\partial x_4}x_1$, $\frac{\partial f_3}{\partial x_5} = 0$, $\frac{\partial f_3}{\partial x_6} = 0$, $\frac{\partial f_3}{\partial x_7} = 0$, $\frac{\partial f_3}{\partial x_8} = 0$, $\frac{\partial f_3}{\partial x_9} = 0$, $\frac{\partial f_3}{\partial x_{10}} = 0$, $\frac{\partial f_3}{\partial x_{11}} = 0$.

where using that $\pi(x_3, x_4) = (a_0 + a_1 x_3 + a_2 x_3^2 + a_3 x_3^3) \times (b_0 + b_1 x_4 + b_2 x_4^2 + b_3 x_4^3)$, it holds that

$$\frac{\partial \pi(x_3, x_4)}{\partial x_3} = (a_1 + 2a_2 x_3 + 3a_3 x_3^2) \times (b_0 + b_1 x_4 + b_2 x_4^2 + b_3 x_4^3)$$

$$\frac{\partial \pi(x_3, x_4)}{\partial x_4} = (a_0 + a_1 x_3 + a_2 x_3^2 + a_3 x_3^3) \times (b_1 + 2b_2 x_4 + 3b_3 x_4^2)$$

Fourth row of the Jacobian matrix $\nabla_x [f(x)] \mid_{(x^*, u^*)}$: $\frac{\partial f_4}{\partial x_1} = 0$, $\frac{\partial f_4}{\partial x_2} = 0$, $\frac{\partial f_4}{\partial x_3} = -\frac{1}{J} \frac{\partial \tau_c(x_3, x_4)}{\partial x_3}$, $\frac{\partial f_4}{\partial x_4} = \frac{1}{J} \frac{\partial \tau_c(x_3, x_4)}{\partial x_3}$, $\frac{\partial f_4}{\partial x_5} = 0$, $\frac{\partial f_4}{\partial x_6} = -\frac{1}{J} \frac{PM}{L_r} x_9$, $\frac{\partial f_4}{\partial x_7} = \frac{1}{J} \frac{PM}{L_r} x_8$, $\frac{\partial f_4}{\partial x_8} = \frac{1}{J} \frac{PM}{L_r} x_7$, $\frac{\partial f_4}{\partial x_9} = -\frac{1}{J} \frac{PM}{L_r} x_6$, $\frac{\partial f_4}{\partial x_{10}} = 0$, $\frac{\partial f_4}{\partial x_{11}} = 0$.

where using that $\tau_c(x_3, x_4) = (c_0 + c_1 x_3 + c_2 x_3^2 + c_3 x_3^3) \times (d_0 + d_1 x_4 + d_2 x_4^2 + d_3 x_4^3)$, it holds that

$$\frac{\partial \tau_c(x_3, x_4)}{\partial x_3} = (c_1 + 2c_2 x_3 + 3c_3 x_3^2) \times (d_0 + d_1 x_4 + d_2 x_4^2 + d_3 x_4^3)$$

$$\frac{\partial \tau_c(x_3, x_4)}{\partial x_4} = (c_0 + c_1 x_3 + c_2 x_3^2 + c_3 x_3^3) \times (d_1 + 2d_2 x_4 + 3d_3 x_4^2)$$

Fifth row of the Jacobian matrix $\nabla_x [f(x)] \mid_{(x^*, u^*)}$: $\frac{\partial f_5}{\partial x_i} = 0$ for $i = 1, 2, \cdots, 11$

Sixth row of the Jacobian matrix $\nabla_x [f(x)] \mid_{(x^*, u^*)}$: $\frac{\partial f_6}{\partial x_1} = 0$, $\frac{\partial f_6}{\partial x_2} = 0$, $\frac{\partial f_6}{\partial x_3} = 0$, $\frac{\partial f_6}{\partial x_4} = ax_3$, $\frac{\partial f_6}{\partial x_5} = 0$, $\frac{\partial f_6}{\partial x_6} = -b$, $\frac{\partial f_6}{\partial x_7} = 0$, $\frac{\partial f_6}{\partial x_8} = aA_r$, $\frac{\partial f_6}{\partial x_9} = ax_4$, $\frac{\partial f_6}{\partial x_{10}} = 0$, $\frac{\partial f_2}{\partial x_{11}} = 0$.

Seventh row of the Jacobian matrix $\nabla_x [f(x)] \mid_{(x^*, u^*)}$: $\frac{\partial f_7}{\partial x_1} = 0$, $\frac{\partial f_7}{\partial x_2} = 0$, $\frac{\partial f_7}{\partial x_3} = 0$, $\frac{\partial f_7}{\partial x_4} = -ax_4$, $\frac{\partial f_7}{\partial x_5} = 0$, $\frac{\partial f_7}{\partial x_6} = 0$, $\frac{\partial f_7}{\partial x_7} = -b$, $\frac{\partial f_8}{\partial x_8} = -ax_4$, $\frac{\partial f_7}{\partial x_9} = aA_r$, $\frac{\partial f_7}{\partial x_{10}} = 0$, $\frac{\partial f_7}{\partial x_{11}} = 0$.

Eighth row of the Jacobian matrix $\nabla_x [f(x)] \mid_{(x^*, u^*)}$: $\frac{\partial f_8}{\partial x_1} = 0$, $\frac{\partial f_8}{\partial x_2} = 0$, $\frac{\partial f_8}{\partial x_3} = 0$, $\frac{\partial f_8}{\partial x_4} = -x_9$, $\frac{\partial f_8}{\partial x_5} = 0$, $\frac{\partial f_8}{\partial x_6} = MA_r$, $\frac{\partial f_8}{\partial x_7} = 0$, $\frac{\partial f_8}{\partial x_8} = -A_r$, $\frac{\partial f_8}{\partial x_9} = -x_4$, $\frac{\partial f_8}{\partial x_{10}} = 0$, $\frac{\partial f_8}{\partial x_{11}} = 0$.

Ninth row of the Jacobian matrix $\nabla_x [f(x)] \mid_{(x^*, u^*)}$: $\frac{\partial f_9}{\partial x_1} = 0$, $\frac{\partial f_9}{\partial x_2} = 0$, $\frac{\partial f_9}{\partial x_3} = 0$, $\frac{\partial f_9}{\partial x_4} = x_8$, $\frac{\partial f_9}{\partial x_5} = 0$, $\frac{\partial f_9}{\partial x_6} = 0$, $\frac{\partial f_9}{\partial x_7} = MA_r$, $\frac{\partial f_9}{\partial x_8} = x_4$, $\frac{\partial f_9}{\partial x_9} = -A_r$, $\frac{\partial f_9}{\partial x_{10}} = 0$, $\frac{\partial f_9}{\partial x_{11}} = 0$.

Tenth row of the Jacobian matrix $\nabla_x [f(x)] \mid_{(x^*, u^*)}$: $\frac{\partial f_{10}}{\partial x_i} = 0$, for $i \neq 10$, and $\frac{\partial f_{10}}{\partial x_{10}} = d$.

Eleventh row of the Jacobian matrix $\nabla_x [f(x)] \mid_{(x^*, u^*)}$: $\frac{\partial f_{11}}{\partial x_i} = 0$, for $i \neq 11$, and $\frac{\partial f_{11}}{\partial x_{11}} = d$.

Computation of the Jacobian matrix $\nabla_x g_1(x) \mid_{(x^*, u^*)}$.

$$\nabla_x g_1(x)\,|_{(x^*,u^*)}=$$

$$=\begin{pmatrix}
-\frac{a_i^2}{2V_i}(P_{atm}-x_1)^{-\frac{1}{2}} & 0 & 0 & 0 & 0 & 0 & 0 & 0 & 0 & 0 & 0 \\
0 & 0 & 0 & 0 & 0 & 0 & 0 & 0 & 0 & 0 & 0 \\
0 & 0 & 0 & 0 & 0 & 0 & 0 & 0 & 0 & 0 & 0 \\
0 & 0 & 0 & 0 & 0 & 0 & 0 & 0 & 0 & 0 & 0 \\
-\frac{1}{2T_r}(P_{atm}-x_1)^{-\frac{1}{2}} & 0 & 0 & 0 & 0 & 0 & 0 & 0 & 0 & 0 & 0 \\
0 & 0 & 0 & 0 & 0 & 0 & 0 & 0 & 0 & 0 & 0 \\
0 & 0 & 0 & 0 & 0 & 0 & 0 & 0 & 0 & 0 & 0 \\
0 & 0 & 0 & 0 & 0 & 0 & 0 & 0 & 0 & 0 & 0 \\
0 & 0 & 0 & 0 & 0 & 0 & 0 & 0 & 0 & 0 & 0 \\
0 & 0 & 0 & 0 & 0 & 0 & 0 & 0 & 0 & 0 & 0 \\
0 & 0 & 0 & 0 & 0 & 0 & 0 & 0 & 0 & 0 & 0
\end{pmatrix} \qquad (3.45)$$

Computation of the Jacobian matrix $\nabla_x g_2(x)\,|_{(x^*,u^*)}$.

$$\nabla_x g_2(x)\,|_{(x^*,u^*)}=$$

$$=\begin{pmatrix}
0 & 0 & 0 & 0 & 0 & 0 & 0 & 0 & 0 & 0 & 0 \\
0 & 0 & 0 & 0 & 0 & 0 & 0 & 0 & 0 & 0 & 0 \\
0 & 0 & 0 & 0 & 0 & 0 & 0 & 0 & 0 & 0 & 0 \\
0 & 0 & 0 & 0 & 0 & 0 & 0 & 0 & 0 & 0 & 0 \\
-\frac{1}{2T_r}(x_2-x_1)^{-\frac{1}{2}} & \frac{1}{2T_r}(x_2-x_1)^{-\frac{1}{2}} & 0 & 0 & 0 & 0 & 0 & 0 & 0 & 0 & 0 \\
0 & 0 & 0 & 0 & 0 & 0 & 0 & 0 & 0 & 0 & 0 \\
0 & 0 & 0 & 0 & 0 & 0 & 0 & 0 & 0 & 0 & 0 \\
0 & 0 & 0 & 0 & 0 & 0 & 0 & 0 & 0 & 0 & 0 \\
0 & 0 & 0 & 0 & 0 & 0 & 0 & 0 & 0 & 0 & 0 \\
0 & 0 & 0 & 0 & 0 & 0 & 0 & 0 & 0 & 0 & 0 \\
0 & 0 & 0 & 0 & 0 & 0 & 0 & 0 & 0 & 0 & 0
\end{pmatrix}$$

$$(3.46)$$

Besides, for the rest of the Jacobian matrices of the columns of the control inputs gain matrix $g(x)$, it holds that

$$\nabla_x g_3(x)\,|_{(x^*,u^*)}=0\in R^{11\times11} \quad \nabla_x g_4(x)\,|_{(x^*,u^*)}=0\in R^{11\times11}$$
$$\nabla_x g_5(x)\,|_{(x^*,u^*)}=0\in R^{11\times11} \quad \nabla_x g_6(x)\,|_{(x^*,u^*)}=0\in R^{11\times11} \qquad (3.47)$$

Thus, after linearization around its current operating point, the model of the gas-compressor which is actuated by the five-phase induction motor is written in the form of Eq (3.41), that is $\dot{x}=Ax+Bu+\tilde{d}$. For the approximately linearized model of the electrically actuated gas-compression systen an H-infinity controller is developed. By denoting the state vector's tracking error as $e=x-x_d$, the controller has the form [206], [195], [198], [204]

$$u(t) = -Ke(t) \tag{3.48}$$

with $K = \frac{1}{r}B^T P$, where P, is a positive definite symmetric matrix, which is obtained from the solution of the Riccati equation

$$A^T P + PA + Q - P(\frac{2}{r}BB^T - \frac{1}{\rho^2}LL^T)P = 0 \tag{3.49}$$

where Q is also a positive semi-definite symmetric matrix, and L is the disturbance inputs gain matrix. The chapter's fault diagnosis method is used in parallel to the control loop of the electrically actuated gas-compression system and without interrupting the system's functioning.

3.2.4   STATE ESTIMATION AND FAULT DIAGNOSIS

3.2.4.1   Robust state estimation with the use of the H_∞ Kalman Filter

Conditioning monitoring has to be implemented with the use of information provided by a small number of sensors and by processing only a small number of state variables. To reconstruct the missing information about the state vector of the system of the gas-compressor which is driven by a five-phase induction motor, it is proposed to use a nonlinear estimation scheme, based on the H-infinity Kalman Filter [195], [198]. By denoting as $A(k)$, $B(k)$ and $C(k)$ the discrete-time equivalents of matrices A, B and C of the linearized state-space model of the system, the recursion of the H_∞ Kalman Filter for the model of the distributed compressors' system, can be formulated in terms of a *measurement update* and a *time update* part

Measurement update:

$$
\begin{aligned}
D(k) &= [I - \theta W(k)P^-(k) + C^T(k)R(k)^{-1}C(k)P^-(k)]^{-1} \\
K(k) &= P^-(k)D(k)C^T(k)R(k)^{-1} \\
\hat{x}(k) &= \hat{x}^-(k) + K(k)[y(k) - C\hat{x}^-(k)]
\end{aligned}
\tag{3.50}
$$

Time update:

$$
\begin{aligned}
\hat{x}^-(k+1) &= A(k)x(k) + B(k)u(k) \\
P^-(k+1) &= A(k)P^-(k)D(k)A^T(k) + Q(k)
\end{aligned}
\tag{3.51}
$$

where it is assumed that the parameter θ is sufficiently small to assure that the covariance matrix $P^-(k)^{-1} - \theta W(k) + C^T(k)R(k)^{-1}C(k)$ will be positive definite. When $\theta = 0$ the H_∞ Kalman Filter becomes equivalent to the standard Kalman Filter. One can measure only a part of the state vector of the system of the five-phase induction motor-driven gas compressor, and can estimate through filtering the rest of the state vector elements which are associated with the turn speed of the compressors or with magnetic flux variables of the stator of the five-phase induction motor.

It is also of worth to mention that the dynamic model of the gas-compressor with actuation from the five-phase induction motor is differentially flat, with flat outputs vector $Y = [x_1, x_4, x_8, x_9, x_{10}, x_{11}]^T$. This allows to consider also flatness-based non-linear Kalman Filters, in place of the chapter's H-infinity Kalman Filter, for treating the state-estimation problem of the electrically actuated gas-compressor [195], [198].

3.2.4.2 Fault detection

The residuals' sequence, that is the differences between (i) the real outputs of the system of the gas-compressor, which is driven by a five-phase induction motor and (ii) the outputs estimated by the H-infinity Kalman Filter (Fig. 3.13) is a discrete error process e_k with dimension $m \times 1$ (here $m = N$ is the dimension of the output measurements vector). Actually, it is a zero-mean Gaussian white-noise process with covariance given by E_k.

A conclusion can be stated based on a measure of certainty that the five-phase induction motor-driven gas-compressor has neither been subjected to a fault nor to a cyberattack. To this end, the following *normalized error square* (NES) is defined [195]

$$\varepsilon_k = e_k^T E_k^{-1} e_k \tag{3.52}$$

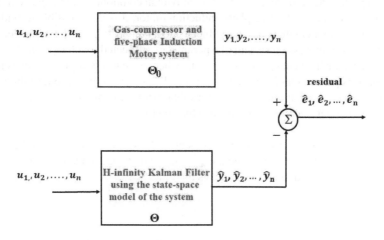

Figure 3.13 Residuals' generation for the gas-compressor and five-phase induction motor system, with the use of Kalman Filtering

The normalized error square follows a χ^2 distribution. An appropriate test for the normalized error sum is to numerically show that the following condition is met within a level of confidence (according to the properties of the χ^2 distribution)

$$E\{\varepsilon_k\} = m \tag{3.53}$$

This can be achieved using statistical hypothesis testing, which is associated with confidence intervals. A 95% confidence interval is frequently applied, which is specified using the probability region $100(1-a)$ with $a = 0.05$. Actually, a two-sided probability region is considered cutting-off two end tails of 2.5% each. For M runs the normalized error square that is obtained is given by

$$\bar{\varepsilon}_k = \frac{1}{M}\sum_{i=1}^{M}\varepsilon_k(i) = \frac{1}{M}\sum_{i=1}^{M}e_k^T(i)E_k^{-1}(i)e_k(i) \tag{3.54}$$

where ε_i stands for the i-th run at time t_k. Then $M\bar{\varepsilon}_k$ will follow a χ^2 density with Mm degrees of freedom. This condition can be checked using a χ^2 test. The hypothesis holds, if the following condition is satisfied

$$\bar{\varepsilon}_k \in [\zeta_1, \zeta_2] \tag{3.55}$$

where ζ_1 and ζ_2 are derived from the tail probabilities of the χ^2 density.

3.2.4.3   Fault isolation

By applying the statistical test into the individual components of the gas-compressor, which is driven by the five-phase induction motor, it is also possible to find out the specific component that has been subjected to a fault [195], [198]. For an electrically-driven gas-compression system of n parameters suspected for change one has to carry out n χ^2 statistical change detection tests, where each test is applied to the subset that comprises parameters $i-1$, i and $i+1$, $i = 1, 2, \cdots, n$. Actually, out of the n χ^2 statistical change detection tests, the ones that exhibit the highest score are those that identify the parameter that has been subjected to change.

In the case of multiple faults one can identify the subset of parameters that has been subjected to change by applying the χ^2 statistical change detection test according to a combinatorial sequence. This means that

$$\binom{n}{k} = \frac{n!}{k!(n-k)!} \tag{3.56}$$

tests have to take place, for all clusters in the model of the five-phase induction motor-driven gas-compressor, that finally comprise n, $n-1$, $n-2$, $\cdots$, 2, 1 parameters. Again the χ^2 tests that give the highest scores indicate the parameters which are most likely to have been subjected to change.

A diagram showing in concise form the parameters of the fault detection and isolation problem for the integrated system of the gas-compressor and five-phase induction motor is shown in Fig. 3.14. These parameters are Δu_1: additive input faults at the five-phase induction motor, Δu_2: additive input faults at the gas-compressor, Δi_{g1}: internal parametric changes at the five-phase induction motor, Δi_{g2}: internal parametric changes at the gas compressor, Δo_{s1}: additive output faults at the five-phase induction motor, Δo_{s2}: additive output faults at the gas compressor, Δi_{s1}: additive faults affecting the inputs of the H-infinity Kalman Filter which are associated with the five-phase induction motor, Δi_{s2}: additive faults affecting the inputs of the H-infinity Kalman Filter which are associated with gas compressor, y_1: the outputs vector of the five-phase induction motor, $\hat{y}_1$: the estimated outputs vector of the five-phase induction motor, which is provided by the H-infinity Kalman Filter, y_2: the outputs vector of the gas compressor, $\hat{y}_2$: the estimated outputs vector of the gas-compressor, which is provided by the H-infinity Kalman Filter, $y_1 - \hat{y}_1$: the residuals vector of the five-phase induction motor, $y_2 - \hat{y}_2$: the residuals vector of the gas compressor

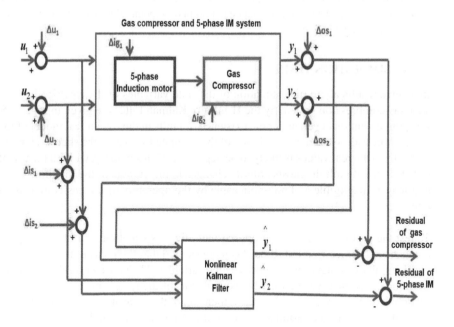

Figure 3.14 Parameters of the fault diagnosis problem for the integrated gas compression system which comprises a centrifugal gas-compressor actuated by a five-phase induction motor

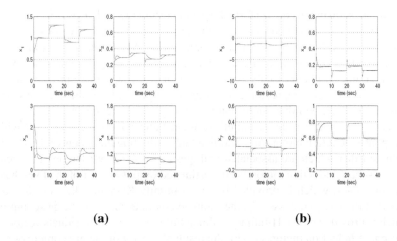

(a)   **(b)**

Figure 3.15 Tracking of a setpoint for the gas compressor, which is driven by a five-phase induction motor (a) convergence of state variables x_1 to x_4 to their reference setpoints (red line: setpoint, blue line: real value, green line: estimated value) and (b) convergence of state variables x_5 to x_8 to their reference setpoints (red line: setpoint, blue line: real value, green line: estimated value)

3.2.5 SIMULATION TESTS

Indicative results about the estimation of the state variables of the electrically actuated gas-compression system by the H-infinity Kalman Filter are given in Fig. 3.15. It is pointed out that even under the existence of faults in the dynamical model of the gas-compression system, the deviation of the estimated values of the state vector elements from the real values is likely to be negligible. Consequently, without the use of statistics-based fault diagnosis criteria, changes in the outputs of the monitored system may be negligible and non-noticeable by the operators of the gas-compression unit.

To test the performance of the proposed fault diagnosis method, simulation experiments have been carried out. It has been considered that the measurement vector of the H-infinity Kalman Filter comprised nine variables. The non-measurable variables from the entire state vector were the components of the rotor's magnetic flux in the five-phase induction motor. Thus, the dimension of the measurements vector, as well as the dimension of the residuals vector of the H-infinity Kalman Filter was equal to $\eta = 9$. This was also the number of the degrees of freedom of the statistical test's χ^2 distribution and also the mean value of the test in the fault-free case. The associated 98% confidence interval was given by the lower and upper bounds $L = 8.60$ and $U = 9.40$, respectively. The concept of the fault diagnosis process has been that as long as the value of the test falls within the above-noted confidence interval no fault has taken place in the gas-compression system. On the contrary, if these thresholds are persistently exceeded by the successive values of the test then it can be concluded that a failure has emerged. By repeating the statistical test in subspaces of the state-space model of the gas-compression system one can also achieve fault isolation. For instance, by applying the statistical test only in the side of the gas-compressor one has a residuals vector of dimension $\eta_1 = 5$ and the confidence interval of the test's χ^2 distribution is $L_1 = 4.70$ and $U_1 = 5.30$. Equivalently, by applying the statistical test only in the side of the five-phase induction motor one has a residuals vector of dimension $\eta_2 = 4$ and the confidence interval of the test's χ^2 distribution is $L_2 = 3.78$ and $U_2 = 4.22$. In the performed tests, the fault was a deviation between 20% and 50% from the nominal value of the affected parameter. To enable detection and isolation of very small parametric changes (incipient faults) the use of a more elaborated χ^2 distribution-based FDI-test has to be considered, as shown in [18-189], [198], [191].

First, fault detection and isolation tests have been performed, in the case that no fault existed in the integrated gas-compression system (neither at the gas-compressor nor at the five-phase induction motor). The associated results are depicted in Fig. 3.16, Fig. 3.17 and Fig. 3.18. It can be noticed that the tests' value was close to the mean value of the χ^2 distribution with 9, 5 and 4 degrees of freedom respectively and remained inside the boundaries defined by the above-noted confidence intervals $[L, U]$, $[L_1, U_1]$ and $[L_2, U_2]$.

Next, fault detection and isolation tests have been performed, in the case that a fault existed at parameter V_i of the gas-compressor. The associated results are depicted in Fig. 3.19, Fig. 3.20 and Fig. 3.21. It can be noticed that the value of the χ^2 test that was performed at the integrated gas-compressor and five-phase IM system exceeded the confidence interval $[L, U]$. Additionally, the value of the χ^2 test that was performed at the gas-compressor exceeded the confidence interval $[L_1, U_1]$. On the contrary, the value of the χ^2 test that was performed at the five-phase IM remained within the confidence interval $[L_2, U_2]$. This has allowed to isolate the failure at the side of the gas-compressor.

Moreover, fault detection and isolation tests have been performed, in the case that a fault existed at parameter V_o of the gas-compressor. The associated results are depicted in Fig. 3.22, Fig. 3.23 and Fig. 3.24. It can be noticed that the value of the χ^2 test that was performed at the integrated gas-compressor and five-phase IM system exceeded the confidence interval $[L, U]$. Additionally, the value of the χ^2 test that was performed at the gas-compressor exceeded the confidence interval $[L_1, U_1]$. On the contrary, the value of the χ^2 test that was performed at the five-phase IM remained within the confidence interval $[L_2, U_2]$. This has allowed to isolate the failure at the side of the gas-compressor.

Furthermore, fault detection and isolation tests have been performed, in the case that a fault existed at parameter T_r of the gas-compressor. The associated results are depicted in Fig. 3.25, Fig. 3.26 and Fig. 3.27. It can be noticed that the value of the χ^2 test that was performed at the integrated gas-compressor and five-phase IM system exceeded the confidence interval $[L, U]$. Additionally, the value of the χ^2 test that was

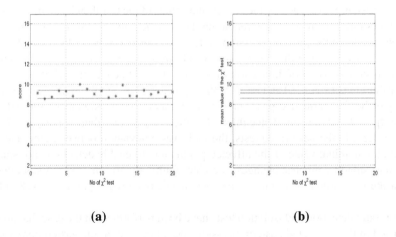

(a) **(b)**

Figure 3.16 FDI test 0 when no fault (parametric change) exists at the gas compressor or at the five-phase induction motor (a) Values of successive χ^2 tests performed at the integrated gas-compressor and five-phase induction motor system and (b) mean value the χ^2 tests performed at the integrated gas-compressor and five-phase induction motor system

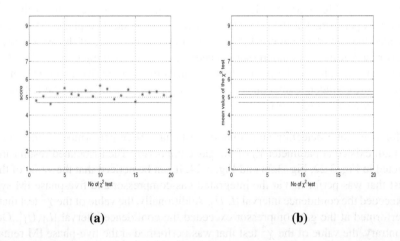

(a) **(b)**

Figure 3.17 FDI test 0 when no fault (parametric change) exists at the gas compressor or at the five-phase induction motor (a) Values of successive χ^2 tests performed at the gas-compressor and (b) mean value the χ^2 tests performed at the gas-compressor

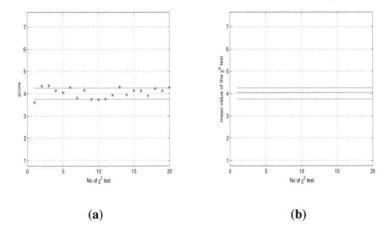

(a)   **(b)**

Figure 3.18 FDI test 0 when no fault (parametric change) exists at the gas compressor or at the five-phase induction motor (a) Values of successive χ^2 tests performed at the five-phase induction motor and (b) mean value the χ^2 tests performed at the five-phase induction motor system

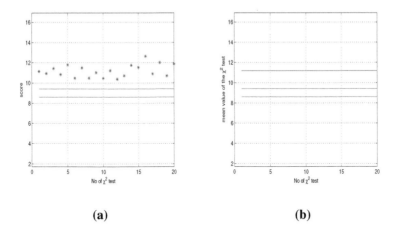

(a)   **(b)**

Figure 3.19 FDI test 1 when a fault (parametric change) exists at parameter V_i of the gas compressor (a) Values of successive χ^2 tests performed at the integrated gas-compressor and five-phase induction motor system and (b) mean value the χ^2 tests performed at the integrated gas-compressor and five-phase induction motor system

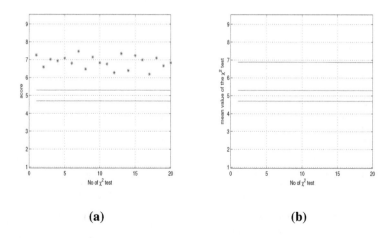

(a)   (b)

Figure 3.20 FDI test 1 when a fault (parametric change) exists at parameter V_i of the gas compressor (a) Values of successive χ^2 tests performed at the gas-compressor and (b) mean value the χ^2 tests performed at the gas-compressor

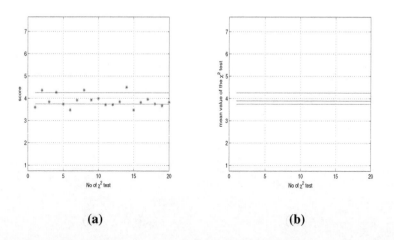

(a)   (b)

Figure 3.21 FDI test 1 when a fault (parametric change) exists at parameter V_i of the gas compressor (a) Values of successive χ^2 tests performed at the five-phase induction motor and (b) mean value the χ^2 tests performed at the five-phase induction motor system

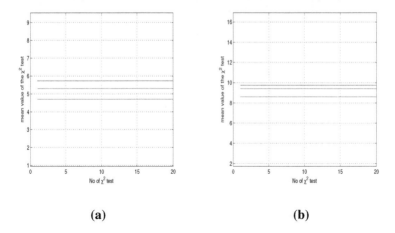

(a) **(b)**

Figure 3.22 FDI test 2 when a fault (parametric change) exists at parameter V_o of the gas compressor (a) Values of successive χ^2 tests performed at the integrated gas-compressor and five-phase induction motor system and (b) mean value the χ^2 tests performed at the integrated gas-compressor and five-phase induction motor system

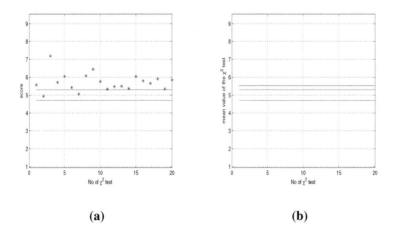

(a) **(b)**

Figure 3.23 FDI test 2 when a fault (parametric change) exists at parameter V_o of the gas compressor (a) Values of successive χ^2 tests performed at the gas-compressor and (b) mean value the χ^2 tests performed at the gas-compressor

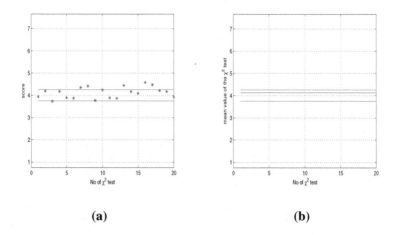

(a)   (b)

Figure 3.24 FDI test 2 when a fault (parametric change) exists at parameter V_o of the gas compressor (a) Values of successive χ^2 tests performed at the five-phase induction motor and (b) mean value the χ^2 tests performed at the five-phase induction motor system

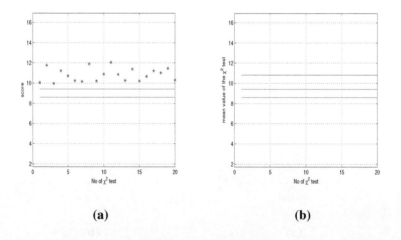

(a)   (b)

Figure 3.25 FDI test 3 when a fault (parametric change) exists at parameter T_r of the gas compressor (a) Values of successive χ^2 tests performed at the integrated gas-compressor and five-phase induction motor system and (b) mean value the χ^2 tests performed at the integrated gas-compressor and five-phase induction motor system

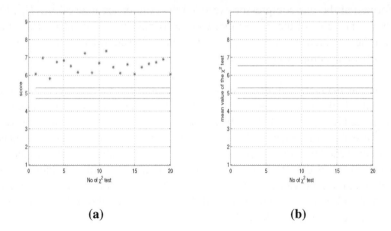

(a)   (b)

Figure 3.26 FDI test 3 when a fault (parametric change) exists at parameter T_r of the gas compressor (a) Values of successive χ^2 tests performed at the gas-compressor and (b) mean value the χ^2 tests performed at the gas-compressor

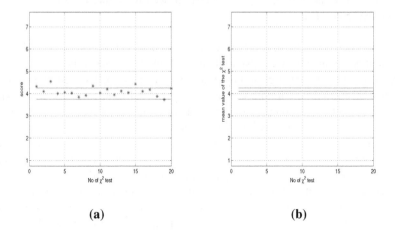

(a)   (b)

Figure 3.27 FDI test 3 when a fault (parametric change) exists at parameter T_r of the gas compressor (a) Values of successive χ^2 tests performed at the five-phase induction motor and (b) mean value the χ^2 tests performed at the five-phase induction motor system.

performed at the gas-compressor exceeded the confidence interval $[L_1, U_1]$. On the contrary, the value of the χ^2 test that was performed at the five-phase IM remained within the confidence interval $[L_2, U_2]$. This has allowed to isolate the failure at the side of the gas-compressor.

In a similar manner, fault detection and isolation tests have been performed, in the case that a fault existed at parameter c of the five-phase IM. The associated results are depicted in Fig. 3.28, Fig. 3.29 and Fig. 3.30. It can be noticed that the value of the χ^2 test that was performed at the integrated gas-compressor and five-phase IM system exceeded the confidence interval $[L, U]$. Additionally, the value of the χ^2 test that was performed at the five-phase IM exceeded the confidence interval $[L_2, U_2]$. On the contrary, the value of the χ^2 test that was performed at the gas-compressor remained within the confidence interval $[L_1, U_1]$. This has allowed to isolate the failure at the side of the five-phase IM.

Finally, fault detection and isolation tests have been performed, in the case that a fault existed at parameter e of the five-phase IM. The associated results are depicted in Fig. 3.31, Fig. 3.32 and Fig. 3.33. It can be noticed that the value of the χ^2 test that was performed at the integrated gas-compressor and five-phase IM system exceeded the confidence interval $[L, U]$. Additionally, the value of the χ^2 test that was performed at the five-phase IM exceeded the confidence interval $[L_2, U_2]$. On the contrary, the value of the χ^2 test that was performed at the gas-compressor remained within the confidence interval $[L_1, U_1]$. This has allowed to isolate the failure at the side of the five-phase IM.

The cumulative results of the fault detection and isolation tests in the electrically actuated gas-compressor are also outlined in Table I. In case of the test that was performed on the integrated gas-compressor and five-phase IM system the confidence interval of normal functioning was $[L, U] = [8.60, 9.40]$. In case of the test that was performed on the gas-compressor the confidence interval of normal functioning was $[L_1, U_1] = [4.70, 5.30]$. In case of the test that was performed on the five-phase IM the confidence interval of normal functioning was $[L_2, U_2] = [3.74, 4.26]$. It can be noticed that the statistical FDI test that was performed on the integrated gas-compressor and five-phase IM system was capable of detecting all faults and of launching a generic alarm about malfunctioning. Moreover, the statistical FDI test that was performed exclusively on the gas-compressor was capable of detecting only faults that had affected the compressor while it gave a no-fault indication when the faults had taken place at the five-phase IM. Equivalently, the statistical FDI test that was performed exclusively on the five-phase IM was capable of detecting only faults that had affected the five-phase IM while it gave a no-fault indication when the faults had

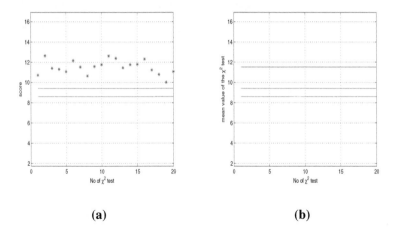

(a) (b)

Figure 3.28 FDI test 4 when a fault (parametric change) exists at parameter c of the gas compressor (a) Values of successive χ^2 tests performed at the integrated gas-compressor and five-phase induction motor system and (b) mean value the χ^2 tests performed at the integrated gas-compressor and five-phase induction motor system

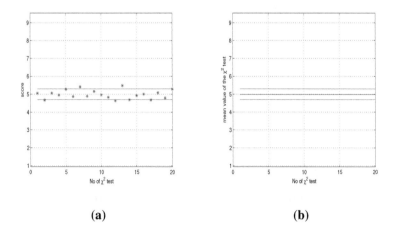

(a) (b)

Figure 3.29 FDI test 4 when a fault (parametric change) exists at parameter c of the gas compressor (a) Values of successive χ^2 tests performed at the gas-compressor and (b) mean value the χ^2 tests performed at the gas-compressor

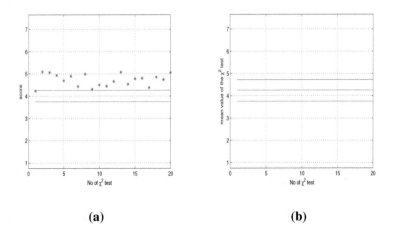

Figure 3.30 FDI test 4 when a fault (parametric change) exists at parameter c of the gas compressor (a) Values of successive χ^2 tests performed at the five-phase induction motor and (b) mean value the χ^2 tests performed at the five-phase induction motor system

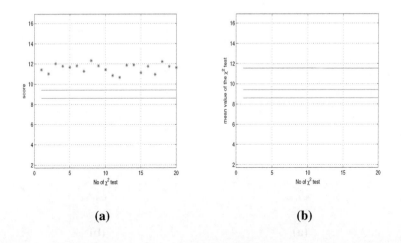

Figure 3.31 FDI test 5 when a fault (parametric change) exists at parameter e of the gas compressor (a) Values of successive χ^2 tests performed at the integrated gas-compressor and five-phase induction motor system and (b) mean value the χ^2 tests performed at the integrated gas-compressor and five-phase induction motor system

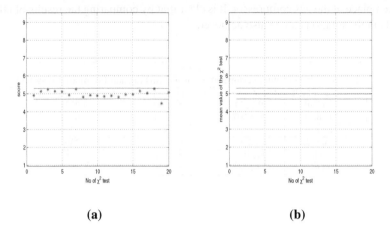

(a)   **(b)**

Figure 3.32 FDI test 5 when a fault (parametric change) exists at parameter *e* of the gas compressor (a) Values of successive χ^2 tests performed at the gas-compressor and (b) mean value the χ^2 tests performed at the gas-compressor

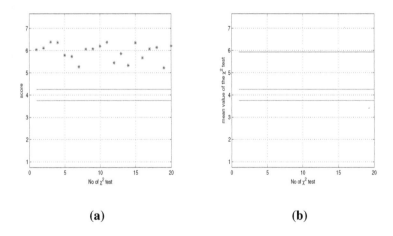

(a)   **(b)**

Figure 3.33 FDI test 5 when a fault (parametric change) exists at parameter *e* of the gas compressor (a) Values of successive χ^2 tests performed at the five-phase induction motor and (b) mean value the χ^2 tests performed at the five-phase induction motor system

taken place at the gas compressor. It is clear that by comparing the results of all three tests fault isolation can be also achieved.

Table I			
FDI tests (mean value) performed at the gas-turbine and five-phase IM system			
type of fault	test at system	test at compressor	test at five-phase IM
no fault	9.2239	5.0590	4.2775
fault V_i	11.8607	6.8284	3.8371
fault V_o	9.7648	5.8405	3.9328
fault T_r	10.2998	6.0630	4.2300
fault c	11.0573	5.2832	4.3287
fault e	11.6982	5.0758	6.1567

4 Fault diagnosis for gas and steam-turbine power generation units

4.1 FAULT DIAGNOSIS FOR THE GAS-TURBINE AND SG ELECTRIC POWER UNIT

4.1.1 OUTLINE

The chapter proposes first a solution for the fault diagnosis problem of a gas-turbine power unit which consists of a synchronous generator that is actuated by a gas-turbine. Gas-turbine power generation units offer a significant part of the electric power that is produced worldwide and their secure and uninterrupted functioning is critical for addressing the growing needs for power supply and for maintaining the stability of the entire power grid. However, due to harsh operating conditions, gas-turbine power units are often subjected to failures in their electromechanical parts (that is the components of the gas-turbines or the components of the electric power generators). Moreover, they can be subjected to cyber-attacks in the form of human intrusion in the power units' control loop. Cyber-attacks target the control software and aim at distorting critical parameters in the power units' functioning. As a consequence of the above, the need for fault diagnosis methods that will be able to detect and isolate incipient faults, and capable of identifying cyber-attacks, has become apparent [199], [201].

As it can be seen in the chapter's modeling part, gas-turbine power generation units exhibit complex dynamics [30], [175]. Thus, condition monitoring of gas-turbine power units is a data-intensive process [152], [144], [265]. A large number of related results are concerned with sensors' fault diagnosis in gas-turbine power generation units [43], [171], [266]. However, the fault diagnosis problem becomes more complicated and exhibits more constraints in the case of faults taking place inside the control loop. To treat the latter case, computational intelligence-based approaches for condition monitoring of gas-turbine power units have been developed [115], [262], [162]. In this context one can note various findings on neural network-based approaches to condition monitoring of gas-turbine power systems [214], [228], [1], [227], [6]. Besides, there exist results on observer-based condition monitoring of gas-turbine power generation units [74], [186], [132], [73]. Additionally, aiming at improved estimation about the monitored gas-turbine power system there have been approaches for fault diagnosis based on Kalman Filtering [138], [139]. The complicated model and the difficulty in measuring specific variables in gas-turbine power generation systems indicate that the conclusive solution of the related

DOI: 10.1201/9781003527657-4

problems of fault tolerant functioning and of fault diagnosis remains a scientific challenge [105], [263], [112], [16], [37], [111].

The chapter proposes a new approach to fault diagnosis and cyber-attacks' detection for electric power generation units comprising a gas-turbine and a synchronous generator. The method relies on a differential flatness theory-based implementation of the Kalman Filter, which is known as Derivative-free nonlinear Kalman Filter. By demonstrating that the dynamic model of the power unit satisfies differential flatness properties, its transformation into an input-output linearized form becomes possible [126], [71], [230], [245], [156]. For the latter description of the system, the solution of the state estimation problem is achieved with a new Kalman Filtering method [195], [75]. This approach is known as Derivative-free nonlinear Kalman Filter and is based on the use of the Kalman Filter's recursion on the linearized equivalent model of the system, as well as on an inverse transformation that provides estimates of the state variables of the initial nonlinear system. The outputs of the monitored gas-turbine power generation unit are compared against the estimated outputs which are provided by the Derivative-free nonlinear Kalman Filter. The differences between these two signals form the residuals' sequence. To diagnose the existence of a failure or cyber-attack the residuals undergo statistical processing [188], [18], [189].

It is shown that the sum of the squares of the residuals' vector weighted by the inverse of the residuals' covariance matrix is a stochastic variable which follows the χ^2 distribution. Next, by using the properties of the aforementioned distribution, confidence intervals are defined which allow for deciding on the power system's condition, with a certainty measure of the order of 96% or 98%. As long as the previous stochastic variable falls within the aforementioned confidence intervals it is concluded that the power unit's function is normal, otherwise it is inferred that the system has undergone a failure or cyber-attack. By implementing the previous procedure with the use of residuals' sub-vectors, that is by performing the statistical test in sub-spaces of the power unit's state-space description, fault isolation can be also performed. Thus, one can identify the component of the power generation unit (gas turbine or generator) which has been subjected to a fault or cyber-attack. Finally, through simulation experiments it has been confirmed that the proposed method is suitable for detecting incipient faults and small parametric changes in the gas-turbine power unit.

4.1.2   DYNAMIC MODEL OF THE GAS-TURBINE POWER UNIT

4.1.2.1   State-space description of the gas-turbine power unit

The power generation unit comprises: (i) the electrical part which is the synchronous power generator [199] and (ii) the thermal power part which is the gas turbine [30].

Dynamics of the synchronous generator: The functioning of the synchronous generator is described in the dq reference frame, which is used for the vectorial representation of the stator's currents and voltages. The rotational motion of the synchronous generator is described by:

$$\dot{\delta} = \omega - \omega_0$$
$$\dot{\omega} = -\frac{D}{2J}(\omega - \omega_0) + \frac{\omega_0}{2J}(P_m - P_e) \tag{4.1}$$

where δ is the turn angle of the generator's rotor, ω is the rotation speed of the rotor with respect to synchronous reference, ω_0 is the synchronous speed of the generator, J is the moment of inertia of the rotor, P_e is the active power of the generator, P_m is the mechanical input torque to the generator which is associated with the mechanical input power, D is the damping constant of the generator and T_e is the electrical torque which is associated to the generated active power. Moreover, the following variables are defined: $\Delta\delta = \delta - \delta_0$ and $\Delta\omega = \omega - \omega_0$ with ω_0 denoting the synchronous speed. The generator's electrical dynamics is described as follows [199]:

$$\dot{E}'_q = \frac{1}{T_{d_o}}(E_f - E_q) \tag{4.2}$$

where E'_q is the quadrature-axis transient voltage of the generator, E_q is the quadrature axis voltage of the generator, T_{d_o} is the direct axis open-circuit transient time constant of the generator and E_f is the equivalent voltage in the excitation coil. The algebraic equations of the synchronous generator are given by

$$E_q = \frac{x_{d\Sigma}}{x'_{d\Sigma}}E'_q - (x_d - x'_d)\frac{V_s}{x_{d\Sigma}}cos(\Delta\delta) \qquad I_q = \frac{V_s}{x_{d\Sigma}}sin(\Delta\delta)$$

$$I_d = \frac{E'_q}{x'_{d\Sigma}} - \frac{V_s}{x'_{d\Sigma}}cos(\Delta\delta) \qquad P_e = \frac{V_s E'_q}{x'_{d\Sigma}}sin(\Delta\delta) \tag{4.3}$$

$$Q_e = \frac{V_s E'_q}{x'_{d\Sigma}}cos(\Delta\delta) - \frac{V_s^2}{x_{d\Sigma}} \qquad V_t = \sqrt{(E'_q - X'_d I_d)^2 + (X'_d I_q)^2}$$

where $x_{d\Sigma} = x_d + x_T + x_L$, $x'_{d\Sigma} = x'_d + x_T + x_L$, x_d is the direct-axis synchronous reactance, x_T is the reactance of the transformer, x'_d is the direct axis transient reactance, x_L is the reactance of the transmission line, I_d and I_q are direct and quadrature axis currents of the generator, V_s is the infinite bus voltage, Q_e is the generator reactive power delivered to the infinite bus, and V_t is the terminal voltage of the generator. Substituting the electrical equations of the synchronous generator given in Eq. (4.81) into the equations of the electrical and mechanical dynamics of the rotor given in Eq. (4.79) and Eq. (4.80) respectively, the complete model of the Single Machine Infinite Bus model is obtained

$$\dot{\delta} = \omega - \omega_0$$
$$\dot{\omega} = -\frac{D}{2J}(\omega - \omega_0) + \omega_0\frac{P_m}{2J} - \omega_0\frac{1}{2J}\frac{V_s E'_q}{x'_{d\Sigma}}sin(\Delta\delta) \tag{4.4}$$
$$\dot{E}'_q = -\frac{1}{T'_d}E'_q + \frac{1}{T_{do}}\frac{x_d - x'_d}{x'_{d\Sigma}}V_s cos(\Delta\delta) + \frac{1}{T_{do}}E_f$$

where $T_d' = \frac{x_{d_\Sigma}'}{x_{d_\Sigma}} T_{do}$ is the time constant of the field winding, and E_f is the excitation voltage. The control input of the generator is E_f.

Dynamics of the gas turbine: The diagram of the gas turbine is depicted in Fig. 4.1. The dynamic model of the gas-turbine is given by:

$$\dot{x} = f(x) + G(x)u$$
$$y = h(x) \tag{4.5}$$

where $x \in R^{5 \times 1}$, $f(x) \in R^{5 \times 1}$, $G(x) \in R^{5 \times 2}$ and $u \in R^{2 \times 1}$. The state vector of the gas turbine is $x = [x_1, x_2, x_3, x_4, x_5]^T$, or equivalently $x = [p_{cc}, T_{cc}, \dot{m}_a, \dot{m}_f, T_{exm}]^T$. The state variables of the model are defined as: p_{cc}: pressure at the combustion chamber, T_{cc}: temperature at the combustion chamber, $\dot{m}_a$: inflow of air in the combustion chamber, $\dot{m}_f$: inflow of fuel in the combustion chamber and T_{exm}: measured temperature at the exhaust [30]. The control inputs vector of the gas turbine are $u = [u_1, u_2]^T = [\dot{m}_a^*, \dot{m}_f^*]^T$, where $\dot{m}_a^*$ is the provided air flow and $\dot{m}_f^*$ is the provided gas flow.

The drift vector $f(x) \in R^{5 \times 1}$ and the control inputs gain matrix $G(x) \in R^{2 \times 1}$ are defined as follows:

$$f(x) = \begin{pmatrix} \frac{R_e}{V_{cc}(c_{pe}R_e)}(x_3 c_{pe} T_a(x_1)) + x_4 H_f - c_{pe} x_2 \dot{m}_e(x_1, x_2)) \\ \frac{x_2 R_e}{x_1} \left\{ \frac{[(R_e - c_{pe})x_2 + c_{pa} T_a(x_1)]x_3 + [(R_e - c_{pe})x_2 + H_f]x_4 - R_e x_2 \dot{m}_e(x_1, x_2)}{V_{cc}(c_{pc} - R_e)} \right\} \\ -\frac{1}{\tau_{igv}} x_3 \\ -\frac{1}{\tau_{at}} x_4 \\ \frac{1}{\tau_{te}}(T_{ex}(x_1, x_2) - x_5) \end{pmatrix} \tag{4.6}$$

$$G(x) = \begin{pmatrix} 0 & 0 \\ 0 & 0 \\ \frac{K_{igv}}{\tau_{igv}} & 0 \\ 0 & \frac{K_{at}}{\tau_{at}} \\ 0 & 0 \end{pmatrix}$$

where

$$T_a(x_1) = \frac{T_{amb}}{n_c}[(\frac{x_1}{P_{amb}})^{\frac{R_a}{C_{pa}}} + n_c - 1] \tag{4.7}$$

$$\dot{m}_e(x_1, x_2) = M_{ao} \frac{\sqrt{T_{ccN}}}{P_{ccN}} \frac{x_1}{\sqrt{x_2}} \tag{4.8}$$

$$T_{ext}(x_1, x_2) = x_2 n_t [(\frac{P_{amb}}{x_1})^{\frac{R_f}{c_{pf}}} + \frac{1}{n_t} - 1] \tag{4.9}$$

Other parameters of the model are: R_e [30]: exhaust gases constant (in J/kg·K), V_{cc}: volume of the combustion chamber, c_{pe}: exhaust specific heat at constant temperature (in J/kg·K), c_{pa}: air specific heat atconstant pressure (in J/kg·K), H_f:

fuel lower heating value: (in J/kg), $\dot{m}_e(x_1,x_2)$: exhaust gas flow (m·kg/sec), τ_{igv}: air intake channel equivalent time constant (in sec), τ_{at}: fuel intake channel equivalent time constant (in sec), τ_{tc}: temperature meter equivalent time constant, K_{igv}: static gain of the air intake channel (in p.u.), K_{at}: static gain of the gain intake channel (in p.u.), a_p: coefficient of the equivalent formulation of the power produced by the gas turbine (in W·(Pa·K)$^\beta$), β: shape function of the equivalent formulation of the power produced by the gas turbine, T_{amb}: ambient temperature, n_c: efficiency factor of the compressor, ρ_{cc}: air density (in kg/m^3).

One can consider as outputs of the gas-turbine model, (i) the power p_{tg} that the gas turbine provides to the synchronous generator and (ii) the temperature of the exhaust gas T_{ext}. These are given by [30]:

$$y = \begin{pmatrix} y_1 \\ y_2 \end{pmatrix} = \begin{pmatrix} p_{tg} \\ T_{exm} \end{pmatrix} = \begin{pmatrix} a_p(x_1 x_2)^\beta \\ x_5 \end{pmatrix} \tag{4.10}$$

Defining the rated power p_N, (in W) coefficient a_p is given by [30]

$$a_p = \frac{p_N}{(\rho_{ccN} T_{ccN})^\beta} \tag{4.11}$$

Joint dynamics of the gas turbine and of the synchronous generator:

The diagram of the electric power generation unit which comprises a gas turbine and a synchronous generator is depicted in Fig. 4.1. The joint state-space model of the synchronous generator and of the gas turbine is written as:

$$\dot{x} = f(x) + G(x)u \tag{4.12}$$

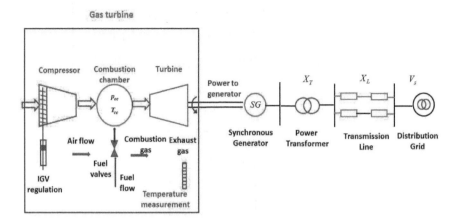

Figure 4.1 Electric power generation unit comprising a gas turbine and a synchronous generator

where the state vector $x \in R^{8 \times 1}$ of the power unit is given by $x = [x_1, x_2, x_3, x_4, x_5, x_6, x_7, x_8]^T$, and is also written explicitly as $x = [\Delta \delta, \delta \omega, E'_q, p_{cc}, T_{cc}, \dot{m}_a, \dot{m}_f, T_{exm}]^T$, with $f(x) \in R^{8 \times 1}$, $G(x) \in R^{8 \times 3}$ and $u \in R^{3 \times 1}$. The control inputs vector of the extended system becomes $u = [u_1, u_2, u_3]^T = [E_f, \dot{m}^*_a, \dot{m}^*_f]^T$. The state-space description of the power unit now becomes:

$$\dot{x}_1 = x_2$$

$$\dot{x}_2 = -\frac{D}{2J}x_2 + \omega_0 \frac{a_p(x_4 x_5)^\beta}{2J} - \frac{\omega_0}{2J}\frac{V_s x_3}{x'_{d\Sigma}}\sin(x_1)$$

$$\dot{x}_3 = -\frac{1}{T'_d}x_3 + \frac{1}{T_{do}}\frac{x_d - x'_d}{x'_{d\Sigma}}V_s\cos(x_1) + \frac{1}{T_{do}}u_1$$

$$\dot{x}_4 = \frac{R_e}{V_{cc}(c_{pe}R_e)}(x_6 c_{pe} T_a(x_1)) + x_7 H_f - c_{pe}x_5\dot{m}_e(x_4, x_5))$$

$$\dot{x}_5 = \frac{x_5 R_e}{x_4}\left\{\frac{[(R_e - c_{pe})x_5 + c_{pa}T_a(x_4)]x_6 + [(R_e - c_{pe})x_5 + H_f]x_7 - R_e x_5 \dot{m}_e(x_4, x_5)}{V_{cc}(c_{pc} - R_e)}\right\}$$

$$\dot{x}_6 = -\frac{1}{\tau_{igv}}x_6 + \frac{K_{igv}}{\tau_{igv}}u_2$$

$$\dot{x}_7 = -\frac{1}{\tau_{at}}x_7 + \frac{K_{at}}{\tau_{at}}u_3$$

$$\dot{x}_8 = \frac{1}{\tau_{te}}(T_{ex}(x_4, x_5) - x_8)$$

(4.13)

According to the above vector $f(x)$ and matrix $G(x)$ appearing in Eq. (4.12) are given by:

$$f(x) = \begin{pmatrix} x_2 \\ -\frac{D}{2J}x_2 + \omega_0 \frac{a_p(x_4 x_5)^\beta}{2J} - \frac{\omega_0}{2J}\frac{V_s x_3}{x'_{d\Sigma}}\sin(x_1) \\ -\frac{1}{T'_d}x_3 + \frac{1}{T_{do}}\frac{x_d - x'_d}{x'_{d\Sigma}}V_s\cos(x_1) \\ \frac{R_e}{V_{cc}(c_{pe}R_e)}(x_6 c_{pe} T_a(x_4)) + x_7 H_f - c_{pe}x_5\dot{m}_e(x_4, x_5)) \\ \frac{x_5 R_e}{x_4}\left\{\frac{[(R_e - c_{pe})x_5 + c_{pa}T_a(x_4)]x_6 + [(R_e - c_{pe})x_5 + H_f]x_7 - R_e x_5 \dot{m}_e(x_4, x_5)}{V_{cc}(c_{pc} - R_e)}\right\} \\ -\frac{1}{\tau_{igv}}x_6 \\ -\frac{1}{\tau_{at}}x_7 \\ \frac{1}{\tau_{te}}(T_{ex}(x_4, x_5) - x_8) \end{pmatrix}$$

(4.14)

$$G(x) = \begin{pmatrix} 0 & 0 & 0 \\ 0 & 0 & 0 \\ \frac{1}{T_{do}} & 0 & 0 \\ 0 & 0 & 0 \\ 0 & 0 & 0 \\ 0 & \frac{K_{igv}}{\tau_{igv}} & 0 \\ 0 & 0 & \frac{K_{at}}{\tau_{at}} \\ 0 & 0 & 0 \end{pmatrix}$$

Figure 4.1 shows the compartments of the gas-turbine and synchronous generator electric power unit. The first part of the power unit is the gas-turbine which provides mechanical torque for the rotation of the generator, through a controlled combustion process. The second part of the power unit is the synchronous generator which is connected to a bus, that is a transmission line of the power grid finally ending at a load that has to be supplied with electric power. The generator receives as input the mechanical torque from the turbine. A secondary input applied to the generator is the electromagnetic torque from the stator's electric field, which is controlled through an excitation voltage on the stator. The generator has to function at the synchronous speed, while faults taking place at either its electrical or its mechanical part can result in loss of synchronization.

4.1.2.2   Faults and cyberattacks affecting the power system

Indicative faults and cyberattacks that may affect the power generation unit, which comprises the gas turbine and the synchronous generator, are shown in Fig. 2. The Kalman Filter's outputs are compared at each sampling instance against the outputs measured from the power generation unit thus providing the residuals' sequence. These faults are: (i) Fault at the control input gains of the gas turbine Δig_1, (ii) Fault at the control input gains of the generator Δig_2, (iii) Fault (additive disturbance) at the control input at the turbine Δu_1, (iv) Fault (additive disturbance) at the control input at the generator Δu_2, (v) Fault at the gas turbine's control input sensors Δis_1, (vi) Fault at the generator's control input sensors, Δis_2, (vii) Fault at the gas turbine's output measurement sensors Δos_1, (viii) Fault at the generator's output measurement sensors Δos_2 and (ix) Cyberattack at the Kalman Filter's output measurements Δof_1.

A fault is considered to be a failure taking place in the electromechanical components that constitute the gas-turbine and the synchronous-generator electric power unit. Such failures are usually caused by the harsh operating conditions of the power unit, perturbations which appear through its connection to the power grid and finally by aging factors. For instance, a failure in the gas supply subsystem of the turbine or in the turbine's shaft is a fault. Equivalently, a failure in the generator's stator or rotor resulting into the machine's malfunctioning is also classified as a fault. On the other side, cyber-attacks are human interventions to the control loop and to the sensors' readings. Consequently, purposely made changes in the parameters of the control or state-estimation software can be classified as cyber-attacks. Besides, distortions to the readings of the sensors used by the control loop can be considered as fault data injection that can be provoked by cyber-attacks. By using redundant filters, the proposed condition monitoring method for the gas-turbine and synchronous generator power unit can distinguish also between malfunctioning of the power unit that is due to faults and malfunctioning of the power unit that is due to cyber-attacks.

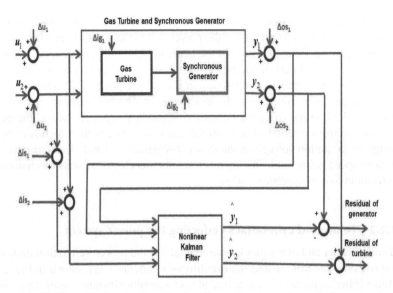

Figure 4.2 Faults and cyberattacks affecting the power generation unit that comprises the gas turbine and the synchronous generator

4.1.3 DIFFERENTIAL FLATNESS-PROPERTIES OF THE GAS-TURBINE POWER UNIT

The computation of setpoints for the functioning of the gas-turbine and synchronous generator power unit is carried out by taking into account the differential flatness properties of this system. The state-space model of the power unit given in Eq. (4.13) is used again:

$$\dot{x}_1 = x_2 \tag{4.15}$$

$$\dot{x}_2 = -\frac{D}{2J}x_2 + \omega_0\frac{a_p(x_4x_5)^\beta}{2J} - \frac{\omega_0}{2J}\frac{V_sx_3}{x'_{d\Sigma}}sin(x_1) \tag{4.16}$$

$$\dot{x}_3 = -\frac{1}{T'_d}x_3 + \frac{1}{T_{do}}\frac{x_d-x'_d}{x'_{d\Sigma}}V_scos(x_1) + \frac{1}{T_{do}}u_1 \tag{4.17}$$

$$\dot{x}_4 = \frac{R_e}{V_{cc}(c_{pe}R_e)}(x_6c_{pe}T_a(x_1)) + x_7H_f - c_{pe}x_5\dot{m}_e(x_4,x_5)) \tag{4.18}$$

$$\dot{x}_5 = \frac{x_5R_e}{x_4}\{\frac{[(R_e-c_{pe})x_5+c_{pa}T_a(x_4)]x_6+[(R_e-c_{pe})x_5+H_f]x_7-R_ex_5\dot{m}_e(x_4,x_5)}{V_{cc}(c_{pc}-R_e)}\}$$

$$\tag{4.19}$$

$$\dot{x}_6 = -\frac{1}{\tau_{igv}} x_6 + \frac{K_{igv}}{\tau_{igv}} u_2 \tag{4.20}$$

$$\dot{x}_7 = -\frac{1}{\tau_{at}} x_7 + \frac{K_{at}}{\tau_{at}} u_3 \tag{4.21}$$

$$\dot{x}_8 = \frac{1}{\tau_{te}} (T_{ex}(x_4, x_5) - x_8) \tag{4.22}$$

The flat outputs of the power unit are taken to be: $Y = [x_1, x_4, x_8]^T$. From Eq. (4.22) of the state-space model one solves with respect to x_5. Thus, it is obtained that;

$$x_5 = \frac{\dot{x}_8 + \frac{1}{\tau_{tc}} x_8}{\frac{R_f}{\frac{1}{\tau_{tc}} n_t [(\frac{P_{amb}}{x_4})^{\frac{R_f}{c_{pf}}} + \frac{1}{n_c} - 1]}} \tag{4.23}$$

From Eq. (4.15), one obtains

$$x_2 = \dot{x}_1 \tag{4.24}$$

thus x_2 is a differential function of the flat outputs of the system. Moreover, after solving Eq. (4.16) with respect to x_3, one obtains

$$x_3 = \frac{\dot{x}_2 - \frac{D}{2J} x_2 - \frac{\omega_0 (x_4 x_5)^\beta}{2J}}{-\frac{\omega_0}{2J} \frac{V_s \sin(x_1)}{x'_{d\Sigma}}} \tag{4.25}$$

From the above relation it can be concluded that x_3 is also a differential function of the flat outputs of the system. Moreover, from Eq. (4.18) and Eq. (4.19), one obtains a system of equations for state variables x_6 and x_7. This system of equations can be written in matrix form as follows:

$$\begin{pmatrix} Rec_{pe} T_a(x_4) & R_e H_f \\ (R_e - c_{pe})x_5 + c_{pe} T_a(x_4) & (R_e - c_{pe})x_5 + H_f \end{pmatrix} \begin{pmatrix} x_6 \\ x_7 \end{pmatrix} =$$
$$= \begin{pmatrix} V_{cc}(c_{pe} - R_e) + R_e(c_{pe} x_5 \dot{m}_e(x_4, x_5)) \\ V_{cc}(c_{pe} - R_e)\frac{x_4 \dot{x}_5}{R_e x_5} + [R_e x_5 + \dot{m}_e(x_4, x_5)] \end{pmatrix} \tag{4.26}$$

After solving Eq. (4.27) with respect to x_6 and x_7, one obtains

$$\begin{pmatrix} x_6 \\ x_7 \end{pmatrix} = \begin{pmatrix} Rec_{pe} T_a(x_4) & R_e H_f \\ (R_e - c_{pe})x_5 + c_{pe} T_a(x_4) & (R_e - c_{pe})x_5 + H_f \end{pmatrix}^{-1} \cdot$$
$$\cdot \begin{pmatrix} V_{cc}(c_{pe} - R_e) + R_e(c_{pe} x_5 \dot{m}_e(x_4, x_5)) \\ V_{cc}(c_{pe} - R_e)\frac{x_4 \dot{x}_5}{R_e x_5} + [R_e x_5 + \dot{m}_e(x_4, x_5)] \end{pmatrix} \tag{4.27}$$

From Eq. (4.27)), one obtains that state variables x_6 and x_7 are also written as differential functions of the flat outputs of the system. Next, by solving Eq. (4.17) with respect to the control input u_1 one gets

$$u_1 = T_{do}[\dot{x}_3 + \frac{1}{T_d'}x_3 - \frac{1}{T_{do}}\frac{x_d - x_d'}{x_{d\Sigma}'}V_s cos(x_1)] \quad (4.28)$$

which signifies that control input u_1 is a differential function of the system's flat outputs. Besides, from Eq. (4.20) and Eq. (4.21) one can solve with respect to the control inputs u_2 and u_3 respectively. Thus, one obtains

$$u_2 = \frac{\tau_{igv}}{K_{igv}}[\dot{x}_6 + \frac{1}{\tau_{igv}}x_6] \quad (4.29)$$

$$u_3 = \frac{\tau_{at}}{K_{at}}[\dot{x}_7 + \frac{1}{\tau_{at}}x_7] \quad (4.30)$$

From Eq. (4.29) and Eq. (4.30) it can be concluded that the control inputs u_2 and u_3 are also differential functions of the flat outputs of the system. As a result of the above, all state variables and the control inputs of the gas-turbine and synchronous generator power unit can be expressed as differential functions of the flat outputs vector. Consequently, the power unit is a differentially flat system.

4.1.4 TRANSFORMATION OF THE POWER UNIT INTO AN INPUT-OUTPUT LINEARIZED FORM

The elements of the flat output vector of the system $y = [x_1, x_4, x_8]^T$ are successively differentiated in time until the control inputs re-appear. From the first row of the state-space model of Eq. (4.13) one has:

$$\dot{x}_1 = x_2 \Rightarrow$$

$$\ddot{x}_1 = \dot{x}_2 = -\frac{D}{2J}x_2 + \omega_0\frac{a_p(x_4 x_5)^\beta}{2J} - \frac{\omega_0}{2J}\frac{V_s x_3}{x_{d\Sigma}'}sin(x_1) \Rightarrow$$

$$x_1^{(3)} = -\frac{D}{2J}\dot{x}_2 + \omega_0\frac{a_p[\beta x_4^{\beta-1}\dot{x}_4 x_5^\beta + \beta x_4^\beta x_5^{\beta-1}\dot{x}_5]}{2J} - \quad (4.31)$$
$$-\frac{\omega_0}{2J}\frac{V_s}{x_{d\Sigma}'}cos(x_1)\dot{x}_1 x_3 - \frac{\omega_0}{2J}\frac{V_s}{x_{d\Sigma}'}sin(x_1)\cdot$$
$$\cdot - \frac{1}{T_d'}x_3 + \frac{1}{T_{do}}\frac{x_d - x_d'}{x_{d\Sigma}'}V_s cos(x_1) + \frac{1}{T_{do}}u_1$$

By regrouping terms, one obtains:

$$x_1^{(3)} = -\frac{D}{2J}\dot{x}_2 + \omega_0\frac{a_p[\beta x_4^{\beta-1}\dot{x}_4 x_5^\beta + \beta x_4^\beta x_5^{\beta-1}\dot{x}_5]}{2J} -$$
$$-\frac{\omega_0}{2J}\frac{V_s}{x_{d\Sigma}'}cos(x_1)\dot{x}_1 x_3 - \frac{\omega_0}{2J}\frac{V_s}{x_{d\Sigma}'}sin(x_1)\cdot \quad (4.32)$$
$$\cdot - [\frac{1}{T_d'}x_3 + \frac{1}{T_{do}}\frac{x_d - x_d'}{x_{d\Sigma}'}V_s cos(x_1)] +$$
$$+ [\frac{\omega_0}{2J}\frac{V_s}{x_{d\Sigma}'}sin(x_1)]\frac{1}{T_{do}}u_1$$

The previous equation can be also written as

$$x_1^{(3)} = f_{a_1}(x) + g_{a_1}(x)u_1 + g_{a_2}(x)u_2 + g_{a_3}(x)u_3 \tag{4.33}$$

where functions $f_{a_1}(x)$ and $g_{a_1}(x)$ are defined as:

$$f_{a_1}(x) = -\frac{D}{2J}\dot{x}_2 + \omega_0 \frac{a_p[\beta x_4^{\beta-1}\dot{x}_4 x_5^{\beta} + \beta x_4^{\beta} x_5^{\beta-1}\dot{x}_5]}{2J} - \\ -\frac{\omega_0}{2J}\frac{V_s}{x_{d\Sigma}}\cos(x_1)\dot{x}_1 x_3 - \frac{\omega_0}{2J}\frac{V_s}{x_{d\Sigma}}\sin(x_1)\cdot \\ \cdots -[\frac{1}{T_d'}x_3 + \frac{1}{T_{do}}\frac{x_d - x_d'}{x_{d\Sigma}}V_s\cos(x_1)] \tag{4.34}$$

$$g_{a_1}(x) = [\frac{\omega_0}{2J}\frac{V_s}{x_{d\Sigma}}\sin(x_1)]\frac{1}{T_{do}} \tag{4.35}$$

while functions $g_{a_2}(x) = 0$ and $g_{a_3}(x) = 0$.

From the fourth row of the state-space model of Eq. (4.13) one has

$$\ddot{x}_4 = \frac{R_e}{V_{cc}(c_{pe}-R_e)}c_{pe}T_a(x_4)\dot{x}_6 + \\ +\frac{R_e}{V_{cc}(c_{pe}-R_e)}c_{pe}\dot{T}_a(x_4)x_6 + \\ +\frac{R_e}{V_{cc}(c_{pe}-R_e)}H_f\dot{x}_7 - \\ -\frac{R_e}{V_{cc}(c_{pe}-R_e)}[c_{pe}\dot{x}_5\dot{m}_a(x_4,x_5) + c_{pe}x_5\dot{m}_e(x_4,x_5)] \tag{4.36}$$

By substituting $\dot{x}_6$ and $\dot{x}_7$ in the previous equation, one obtains:

$$\ddot{x}_4 = \frac{R_e}{V_{cc}(c_{pe}-R_e)}c_{pe}T_a(x_4)[-\frac{1}{\tau_{igv}}x_6 + \frac{K_{igv}}{\tau_{igv}}u_2] + \\ +\frac{R_e}{V_{cc}(c_{pe}-R_e)}c_{pe}\dot{T}_a(x_4)x_6 + \\ +\frac{R_e}{V_{cc}(c_{pe}-R_e)}H_f[\frac{1}{\tau_{at}}x_7 + \frac{K_{at}}{\tau_{at}}u_3] - \\ -\frac{R_e}{V_{cc}(c_{pe}-R_e)}[c_{pe}\dot{x}_5\dot{m}_a(x_4,x_5) + c_{pe}x_5\dot{m}_e(x_4,x_5)] \tag{4.37}$$

Thus, by regrouping terms one arrives at:

$$\ddot{x}_4 = \frac{R_e}{V_{cc}(c_{pe}-R_e)}c_{pe}T_a(x_4)[-\frac{1}{\tau_{igv}}x_6] + \\ +\frac{R_e}{V_{cc}(c_{pe}-R_e)}c_{pe}\dot{T}_a(x_4)x_6 + \\ +\frac{R_e}{V_{cc}(c_{pe}-R_e)}H_f[-\frac{1}{\tau_{at}}x_7] - \\ -\frac{R_e}{V_{cc}(c_{pe}-R_e)}[c_{pe}\dot{x}_5\dot{m}_a(x_4,x_5) + c_{pe}x_5\dot{m}_e(x_4,x_5)] \\ +\{\frac{R_e}{V_{cc}(c_{pe}-R_e)}c_{pe}T_a(x_4)\}\frac{K_{igv}}{\tau_{igv}}u_2 + \\ +\{\frac{R_e}{V_{cc}(c_{pe}-R_e)}H_f\frac{K_{at}}{\tau_{at}}\}u_3 \tag{4.38}$$

The above equation can be written in the compact form

$$\ddot{x}_4 = f_{b_1}(x) + g_{b_1}(x)u_1 + g_{b_2}(x)u_2 + g_{b_3}(x)u_3 \tag{4.39}$$

where functions $f_{b_1}(x)$, $g_{b_2}(x)$ and $g_{b_3}(x)$ are defined as

$$
\begin{aligned}
f_{b_1}(x) = {} & \frac{R_e}{V_{cc}(c_{pe}-R_e)}c_{pe}T_a(x_4)[-\frac{1}{\tau_{igv}}x_6] + \\
& + \frac{R_e}{V_{cc}(c_{pe}-R_e)}c_{pe}\dot{T}_a(x_4)x_6 + \\
& + \frac{R_e}{V_{cc}(c_{pe}-R_e)}H_f[-\frac{1}{\tau_{at}}x_7] - \\
& - \frac{R_e}{V_{cc}(c_{pe}-R_e)}[c_{pe}\dot{x}_5\dot{m}_a(x_4,x_5)+c_{pe}x_5\ddot{m}_e(x_4,x_5)]
\end{aligned}
\tag{4.40}
$$

$$
g_{b_2}(x) = \{\frac{R_e}{V_{cc}(c_{pe}-R_e)}c_{pe}T_a(x_4)\}\frac{K_{igv}}{\tau_{igv}}
\tag{4.41}
$$

$$
g_{b_3}(x) = \{\frac{R_e}{V_{cc}(c_{pe}-R_e)}H_f\}\frac{K_{at}}{\tau_{at}}
\tag{4.42}
$$

while function $g_{b_1}(x) = 0$.

From the eighth row of the state-space model of Eq. (4.13) one has

$$
\begin{aligned}
\dot{x}_8 = {} & \frac{1}{\tau_{tc}}x_5 n_t[(\frac{P_{amb}}{x_4})^{\frac{R_f}{c_{pf}}}+\frac{1}{n_t}-1]-\frac{1}{\tau_{tc}}x_8 \Rightarrow \\
\ddot{x}_8 = {} & \frac{1}{\tau_{tc}}\dot{x}_5 n_t[(\frac{P_{amb}}{x_4})^{\frac{R_f}{c_{pf}}}+\frac{1}{n_t}-1] + \\
& + \frac{1}{\tau_{tc}}x_5[\frac{R_f}{c_{pf}}(\frac{P_{amb}}{x_4})^{\frac{R_f}{c_{pf}}-1}\cdot(-\frac{P_{amb}}{x_4^2}\dot{x}_4] - \\
& - \frac{1}{\tau_{tc}}\dot{x}_8
\end{aligned}
\tag{4.43}
$$

Next, the terms $\dot{x}_4$, $\dot{x}_5$ and $\dot{x}_8$ are expressed as functions of x_6 and x_7. Thus one can write

$$
\dot{x}_4 = a_4(x)x_6 + b_4(x)x_7 + c_4(x)
\tag{4.44}
$$

$$
\dot{x}_5 = a_5(x)x_6 + b_5(x)x_7 + c_5(x)
\tag{4.45}
$$

$$
\dot{x}_8 = a_8(x)x_6 + b_8(x)x_7 + c_8(x)
\tag{4.46}
$$

$a_4(x) = \frac{R_e}{V_{cc}(c_{pe}-R_e)}(c_{pe}T_a(x_1))$, $b_4(x) = \frac{R_e}{V_{cc}(c_{pe}-R_e)}H_f$, $c_4(x) = -\frac{R_e c_{pe}x_5}{V_{cc}(c_{pe}-R_e)}\cdot\dot{m}_e(x_4,x_5)$, $a_5(x) = \frac{x_5 R_e}{x_4}(R_e-c_{pe})x_5+c_{pa}T_a(x_4)]$, $b_5(x) = \frac{x_5 R_e}{x_4}[(R_e-c_{pe})x_5+H_f]$, $c_5(x) = \frac{x_5 R_e}{x_4}[-R_e x_5\dot{m}_e(x_4,x_5)]$, $a_8(x) = 0$, $b_9(x) = 0$ and $c_8(x) = 0$.

By substituting $\dot{x}_4$, $\dot{x}_5$ and $\dot{x}_8$ into Eq. (4.43), one obtains

$$
\begin{aligned}
\ddot{x}_8 = {} & \frac{1}{\tau_{tc}}n_t[(\frac{P_{amb}}{x_4})^{\frac{R_f}{c_{pf}}}+\frac{1}{n_t}-1][a_5(x)x_6+b_5(x)x_7+c_5(x)] + \\
& + \frac{1}{\tau_{tc}}x_5[\frac{R_f}{c_{pf}}(\frac{P_{amb}}{x_4})^{\frac{R_f}{c_{pf}}-1}\cdot(-\frac{P_{amb}}{x_4^2})[a_4(x)x_6+b_4(x)x_7+c_4(x)] - \\
& - \frac{1}{\tau_{tc}}c_8(x)
\end{aligned}
\tag{4.47}
$$

Moreover, by defining

$$q_1(x) = \frac{1}{\tau_{tc}} n_t [(\frac{P_{amb}}{x_4})^{\frac{R_f}{c_{pf}}} + \frac{1}{n_t} - 1]$$

$$q_2(x) = \frac{1}{\tau_{tc}} x_5 [\frac{R_f}{c_{pf}} (\frac{P_{amb}}{x_4})^{\frac{R_f}{c_{pf}} - 1} \cdot (-\frac{P_{amb}}{x_4^2})]$$

Eq. (4.47) can be written as:

$$\ddot{x}_8 = q_1(x)a_5(x)x_6 + q_1(x)b_5(x)x_7 + q_1(x)c_5(x) +$$
$$+ q_2(x)a_4(x)x_6 + q_2(x)b_4(x)x_7 + q_2(x)c_4(x) - \tag{4.48}$$
$$- \frac{1}{\tau_{tc}} c_8(x)$$

By regrouping terms in the previous equation one has

$$\ddot{x}_8 = [q_1(x)a_5(x) + q_2(x)a_4(x)]x_6 +$$
$$[q_1(x)b_5(x) + q_2(x)b_4(x)]x_7 + \tag{4.49}$$
$$[q_1(x)c_5(x) + q_2(x)c_4(x) - \frac{1}{\tau_{tc}} c_8(x)]$$

Next, by denoting the functions $m_1(x) = [q_1(x)a_5(x) + q_2(x)a_4(x)]$, $m_2(x) = [q_1(x)b_5(x) + q_2(x)b_4(x)]$ and $m_3(x) = [q_1(x)c_5(x) + q_2(x)c_4(x) - \frac{1}{\tau_{tc}} c_8(x)]$, Eq. (4.49) is written as

$$\ddot{x}_8 = m_1(x)x_6 + m_2(x)x_7 + m_3(x) \tag{4.50}$$

By differentiating once more in time Eq. (4.50) one gets

$$x_8^{(3)} = \dot{m}_1(x)x_6 + m_1(x)\dot{x}_6 + \dot{m}_2(x)x_7 + m_2(x)\dot{x}_7 + \dot{m}_3(x) \tag{4.51}$$

or equivalently

$$x_8^{(3)} = m_1(x)\dot{x}_6 + m_2(x)\dot{x}_7 + [\dot{m}_1(x)x_6 + \dot{m}_2(x)x_7 + \dot{m}_3(x)] \tag{4.52}$$

Next, by substituting in Eq. (4.52) the terms $\dot{x}_6$ and $\dot{x}_7$, one obtains

$$x_8^{(3)} = m_1(x)[-\frac{1}{\tau_{igv}} x_6 + \frac{K_{igv}}{\tau_{igv}} u_2] +$$
$$+ m_2(x)[-\frac{1}{\tau_{at}} x_7 + \frac{K_{at}}{\tau_{at}} u_3] + \tag{4.53}$$
$$+ [\dot{m}_1(x)x_6 + \dot{m}_2(x)x_7 + \dot{m}_3(x)]$$

By regrouping terms in Eq. (4.53), one obtains

$$x_8^{(3)} = m_1(x)[-\frac{1}{\tau_{igv}} x_6] + m_2(x)[-\frac{1}{\tau_{at}} x_7 +] + [\dot{m}_1(x)x_6 + \dot{m}_2(x)x_7 + \dot{m}_3(x)] +$$
$$+ m_1(x)\frac{K_{igv}}{\tau_{igv}} u_2 + m_2(x)\frac{K_{at}}{\tau_{at}} u_3$$
$$\tag{4.54}$$

The above relation can be also written in the following form:

$$x_8^{(3)} = f_{c_1}(x) + g_{c_1}(x)u_1 + g_{c_2}(x)u_2 + g_{c_3}(x)u_3 \tag{4.55}$$

where functions $f_{c_1}(x)$, $g_{c_2}(x)$, and $g_{c_3}(x)$ are defined as

$$f_{c_1}(x) = m_1(x)[-\frac{1}{\tau_{igv}}x_6] + m_2(x)[-\frac{1}{\tau_{at}}x_7+] + [\dot{m}_1(x)x_6 + \dot{m}_2(x)x_7 + \dot{m}_3(x)] \quad (4.56)$$

$$g_{c_2}(x) = m_1(x)\frac{K_{igv}}{\tau_{igv}} \quad (4.57)$$

$$g_{c_3}(x) = m_2(x)\frac{K_{at}}{\tau_{at}} \quad (4.58)$$

while one also has that $g_{c_1}(x) = 0$. Consequently, the gas-turbine and synchronous generator electric power system can be written in the following input-output linearized form:

$$\begin{pmatrix} x_1^{(3)} \\ \ddot{x}_4 \\ x_8^{(3)} \end{pmatrix} = \begin{pmatrix} f_a(x) \\ f_b(x) \\ f_c(x) \end{pmatrix} + \begin{pmatrix} g_{a_1}(x) & g_{a_2}(x) & g_{a_3}(x) \\ g_{b_1}(x) & g_{b_2}(x) & g_{b_3}(x) \\ g_{c_1}(x) & g_{c_2}(x) & g_{c_3}(x) \end{pmatrix} \begin{pmatrix} u_1 \\ u_2 \\ u_3 \end{pmatrix} \quad (4.59)$$

Using the previous relation one can denote the drift matrix $F(x)$ and the input gains matrix $G(x)$ as follows:

$$F(x) = \begin{pmatrix} f_a(x) \\ f_b(x) \\ f_c(x) \end{pmatrix} \quad G(x) = \begin{pmatrix} g_{a_1}(x) & g_{a_2}(x) & g_{a_3}(x) \\ g_{b_1}(x) & g_{b_2}(x) & g_{b_3}(x) \\ g_{c_1}(x) & g_{c_2}(x) & g_{c_3}(x) \end{pmatrix} \quad (4.60)$$

Using the input-output linearized form of the gas-turbine and synchronous-generator power unit, the stabilizing feedback controller is designed as follows:

$$\begin{pmatrix} u_1 \\ u_2 \\ u_3 \end{pmatrix} = \begin{pmatrix} g_{a_1}(x) & g_{a_2}(x) & g_{a_3}(x) \\ g_{b_1}(x) & g_{b_2}(x) & g_{b_3}(x) \\ g_{c_1}(x) & g_{c_2}(x) & g_{c_3}(x) \end{pmatrix}^{-1} \cdot$$

$$\cdot \left[\begin{pmatrix} x_1^{(3),d} \\ \ddot{x}_4^d \\ x_8^{(3),d} \end{pmatrix} - \begin{pmatrix} f_{a_1}(x) \\ f_{b_1}(x) \\ f_{c_1}(x) \end{pmatrix} + \begin{pmatrix} -k_{11}(\ddot{x}_1 - \ddot{x}_1^d) - k_{12}(\dot{x}_1 - \dot{x}_1^d) - k_{13}(x_1 - x_1^d) \\ -k_{21}(\dot{x}_4 - \dot{x}_4^d) - k_{22}(x_4 - x_4^d) \\ -k_{31}(\ddot{x}_8 - \ddot{x}_8^d) - k_{32}(\dot{x}_8 - \dot{x}_8^d) - k_{33}(x_8 - x_8^d) \end{pmatrix} \right]$$

$$\hspace{11cm} (4.61)$$

The aforementioned control law results into the following tracking error dynamics

$$(x_1^{(3)} - x_1^{(3),d}) + k_{11}(\ddot{x}_1 - \ddot{x}_1^d) + k_{12}(\dot{x}_1 - \dot{x}_1^d) + k_{13}(x_1 - x_1^d) = 0$$

$$(\ddot{x}_4 - \ddot{x}_4^d) + k_{21}(\dot{x}_4 - \dot{x}_4^d) + k_{22}(x_4 - x_4^d) = 0 \quad (4.62)$$

$$(x_8^{(3)} - x_8^{(3),d}) + k_{31}(\ddot{x}_8 - \ddot{x}_8^d) + k_{32}(\dot{x}_8 - \dot{x}_8^d) + k_{33}(x_8 - x_8^d) = 0$$

and by defining the tracking error variables $e_1 = x_1 - x_1^d$, $e_4 = x_4 - x_4^d$ and $e_8 = x_8 - x_8^d$, the tracking error dynamics is written as

$$e_1^{(3)} + k_{11}\ddot{e}_1 + k_{12}\dot{e}_1 + k_{13}e_1 = 0$$
$$\ddot{e}_4 + k_{21}\dot{e}_4 + k_{22}e_4 = 0 \tag{4.63}$$
$$e_8^{(3)} + k_{31}\ddot{e}_8 + k_{32}\dot{e}_8 + k_{33}e_1 = 0$$

By selecting the previously defined feedback control gains so as the characteristic polynomials which are associated with the previous tracking error dynamics to be Hurwitz stable, one has that the tracking errors are asymptotically eliminated. This means

$$lim_{t \to \infty} e_1(t) = 0 \quad lim_{t \to \infty} e_2(t) = 0 \quad lim_{t \to \infty} e_3(t) = 0 \tag{4.64}$$

Consequently, using flatness-based control, one assures the global asymptotic stability of the control loop that comprises the gas turbine and the synchronous generator.

4.1.5   STATE ESTIMATION USING THE DERIVATIVE-FREE NONLINEAR KALMAN FILTER

The input-output linearized model of the gas-turbine and synchronous generator model is:

$$\begin{pmatrix} x_1^{(3)} \\ \ddot{x}4 \\ x_8^{(3)} \end{pmatrix} = \begin{pmatrix} f_{a_1} \\ f_{b_1} \\ f_{c_1} \end{pmatrix} + \begin{pmatrix} g_{a_1}(x) & g_{a_2}(x) & g_{a_3}(x) \\ g_{b_1}(x) & g_{b_2}(x) & g_{b_3}(x) \\ g_{c_1}(x) & g_{c_2}(x) & g_{c_3}(x) \end{pmatrix} \begin{pmatrix} u_1 \\ u_2 \\ u_3 \end{pmatrix} \tag{4.65}$$

Equivalently the dynamics of the model can be described by:

$$x_1^{(3)} = f_{a_1}(x) + g_{a_1}(x)u_1 + g_{a_2}(x)u_2 + g_{a_3}(x)u_3$$
$$\ddot{x}_4 = f_{b_1}(x) + g_{b_1}(x)u_1 + g_{b_2}(x)u_2 + g_{b_3}(x)u_3 \tag{4.66}$$
$$x_8^{(3)} = f_{c_1}(x) + g_{c_1}(x)u_1 + g_{c_2}(x)u_2 + g_{c_3}(x)u_3$$

Next, the following transformed control inputs are defined

$$v_1 = f_{a_1}(x) + g_{a_1}(x)u_1 + g_{a_2}(x)u_2 + g_{a_3}(x)u_3$$
$$v_2 = f_{b_1}(x) + g_{b_1}(x)u_1 + g_{b_2}(x)u_2 + g_{b_3}(x)u_3 \tag{4.67}$$
$$v_3 = f_{c_1}(x) + g_{c_1}(x)u_1 + g_{c_2}(x)u_2 + g_{c_3}(x)u_3$$

and the state-space description of the system becomes

$$x_1^{(3)} = v_1 \quad \ddot{x}_4 = v_2 \quad x_8^{(3)} = v_3 \tag{4.68}$$

Next, the state-variables of the gas-turbine and synchronous generator are redefined as follows: $z_1 = x_1$, $z_2 = \dot{x}_1$, $z_3 = \ddot{x}_1$, $z_4 = x_4$, $z_5 = \dot{x}_4$, $z_6 = x_8$, $z_7 = \dot{x}_8$, and $z_8 = \ddot{x}_8$.

The system's state vector allows to rewrite the state-space model of the system in the following linear canonical (Brunovksy) form:

$$
\begin{pmatrix} \dot{z}_1 \\ \dot{z}_2 \\ \dot{z}_3 \\ \dot{z}_4 \\ \dot{z}_5 \\ \dot{z}_6 \\ \dot{z}_7 \\ \dot{z}_8 \end{pmatrix} =
\begin{pmatrix}
0\,1\,0\,0\,0\,0\,0\,0 \\
0\,0\,1\,0\,0\,0\,0\,0 \\
0\,0\,0\,0\,0\,0\,0\,0 \\
0\,0\,0\,0\,1\,0\,0\,0 \\
0\,0\,0\,0\,0\,0\,0\,0 \\
0\,0\,0\,0\,0\,0\,1\,0 \\
0\,0\,0\,0\,0\,0\,0\,1 \\
0\,0\,0\,0\,0\,0\,0\,0
\end{pmatrix}
\begin{pmatrix} z_1 \\ z_2 \\ z_3 \\ z_4 \\ z_5 \\ z_6 \\ z_7 \\ z_8 \end{pmatrix} +
\begin{pmatrix}
0\,0\,0 \\
0\,0\,0 \\
1\,0\,0 \\
0\,0\,0 \\
0\,1\,0 \\
0\,0\,0 \\
0\,0\,0 \\
0\,0\,1
\end{pmatrix}
\begin{pmatrix} v_1 \\ v_2 \\ v_3 \end{pmatrix} \quad
z_m = \begin{pmatrix}
1\,0\,0\,0\,0\,0\,0\,0 \\
0\,0\,0\,1\,0\,0\,0\,0 \\
0\,0\,0\,0\,0\,1\,0\,0
\end{pmatrix} z
$$

(4.69)

where the state vector is defined as $z = [z_1, z_2, z_3, z_4, z_5, z_6, z_7, z_8]^T$, z_m is the associated measurements vector, and the inputs vector is defined as $v = [u_1, u_2, u_3]^T$. The previous description of the system is written in the concise state-space form

$$
\begin{aligned}
\dot{z} &= Az + B\tilde{v} \\
z_m &= Cz
\end{aligned}
$$

(4.70)

To perform estimation of the non-measurable state-vector elements, the following state-observer is defined:

$$
\begin{aligned}
\dot{\hat{z}} &= A_o\hat{z} + B_o v + K_f(z_m - \hat{z}_m) \\
\hat{z}_m &= C_o\hat{z}
\end{aligned}
$$

(4.71)

where $v = [v_1, v_2, v_3]^T$, $A_o = A$, $B_0 = B$ ad $C_o = C$. The observer's gain is computed through the Kalman Filter's recursion. The application of Kalman Filtering to the linearized equivalent description of the power generation unit is the Derivative-free nonlinear Kalman Filter . Matrices A_o, B_o and C_o are discretized using common discretization methods. This provides matrices A_d, B_d and C_d. Matrices Q and R denote the process and measurement noise covariance matrices. Matrix P is the state-vector error covariance matrix. The use of the Derivative-free nonlinear Kalman Filter as a state estimator is shown in Fig. 4.3.

The stages of the Kalman Filter are:

measurement update:

$$
\begin{aligned}
K_f(k) &= P^-(k)C_d^T[C_dP^-(k)C_d^T + R]^{-1} \\
\hat{z}(k) &= \hat{z}^-(k) + K_f(k)[z_m - \hat{z}_m] \\
P(k) &= P^-(k) - K_f(k)C_dP^-(k)
\end{aligned}
$$

(4.72)

time update:

$$
\begin{aligned}
P^-(k+1) &= A_dP(k)A_d^T + Q \\
\hat{z}^-(k+1) &= A_d\hat{z}(k) + B_dv(k)
\end{aligned}
$$

(4.73)

4.1.6 STATISTICAL FAULT DIAGNOSIS USING THE KALMAN FILTER

4.1.6.1 Fault detection

The residuals' sequence, that is the differences between (i) the real outputs of the gas-turbine power unit and (ii) the outputs estimated by the Kalman Filter (Fig. 4.4) is a discrete error process e_k with dimension $m \times 1$ (here $m = N$ is the dimension of the output measurements vector). Actually, it is a zero-mean Gaussian white-noise process with covariance given by E_k.

A conclusion can be stated based on a measure of certainty that the power unit has neither been subjected to a fault nor to a cyberattack. To this end, the following *normalized error square* (NES) is defined [195]

$$\varepsilon_k = e_k^T E_k^{-1} e_k \qquad (4.74)$$

The sum of this normalized residuals' square follows a χ^2 distribution, with a number of degrees of freedom that is equal to the dimension of the residuals' vector. An appropriate test for the normalized error sum is to numerically show that the following condition is met within a level of confidence (according to the properties of the χ^2 distribution)

$$E\{\varepsilon_k\} = m \qquad (4.75)$$

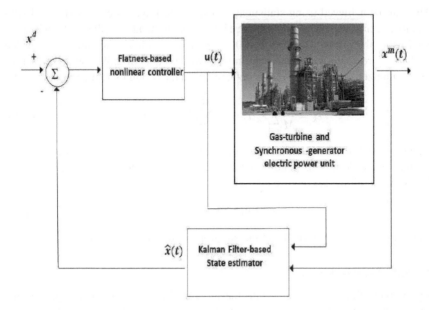

Figure 4.3 Use of the Derivative-free nonlinear Kalman Filter as a state estimator in the flatness-based control loop of the gas turbine and synchronous generator power unit

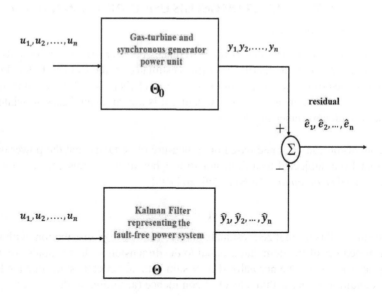

Figure 4.4 Residuals' generation for the gas-turbine power unit, with the use of Kalman Filtering

This can be achieved using statistical hypothesis testing, which is associated with confidence intervals. A 95% confidence interval is frequently applied, which is specified using the probability region $100(1 - a)$ with $a = 0.05$. Actually, a two-sided probability region is considered cutting-off two end tails of 2.5% each. For M runs the normalized error square, that is obtained from the residuals sequence, is given by

$$\bar{\varepsilon}_k = \frac{1}{M}\sum_{i=1}^{M}\varepsilon_k(i) = \frac{1}{M}\sum_{i=1}^{M}e_k^T(i)E_k^{-1}(i)e_k(i) \tag{4.76}$$

where ε_i stands for the i-th run at time t_k. Then $M\bar{\varepsilon}_k$ will follow a χ^2 density with Mm degrees of freedom. This condition can be checked using a χ^2 test. The hypothesis holds, if the following condition is satisfied

$$\bar{\varepsilon}_k \in [\zeta_1, \zeta_2] \tag{4.77}$$

where ζ_1 and ζ_2 are derived from the tail probabilities of the χ^2 density.

4.1.6.2 Fault isolation

By applying the statistical test into the individual components of the gas-turbine and synchronous generator power unit, it is also possible to find out the specific component that has been subjected to a fault or cyber-attack [195], [198]. For a gas-turbine power unit of n parameters suspected for change one has to carry out n χ^2

statistical change detection tests, where each test is applied to the subset that comprises parameters $i-1$, i and $i+1$, $i = 1, 2, \cdots, n$. Actually, out of the n χ^2 statistical change detection tests, the ones that exhibit the highest score are those that identify the parameter that has been subject to change.

In the case of multiple faults one can identify the subset of parameters that has been subjected to change by applying the χ^2 statistical change detection test according to a combinatorial sequence. This means that

$$\binom{n}{k} = \frac{n!}{k!(n-k)!} \tag{4.78}$$

tests have to take place, for all clusters in the gas-turbine power unit, that finally comprise n, $n-1$, $n-2$, $\cdots$, 2, 1 parameters. Again the χ^2 tests that give the highest scores indicate the parameters which are most likely to have been subjected to change.

As a whole, the concept of the proposed fault detection and isolation method is again simple. The sum of the squares of the residuals' vector, weighted by the inverse of the residuals' covariance matrix, stands for a stochastic variable which follows the χ^2 distribution. Actually, this is a multi-dimensional χ^2 distribution and the number of its degrees of freedom is equal to the dimension of the residuals' vector. Since, there are 3 measurable outputs of the gas-turbine power unit the residuals' vector is of dimension 3 and the number of degrees of freedom is also 3. Next, from the properties of the χ^2 distribution, the mean value of the aforementioned stochastic variable in the fault-free case should be also 3. However, due to having sensor measurements subject to noise, the value of the statistical test in the fault-free case will not be precisely equal to 3 but it may vary within a small range around this value. This range is determined by the confidence intervals of the χ^2 distribution. For a probability of 98% to get a value of the stochastic variable about 3, the associated confidence interval is given by the lower bound $L = 2.87$ and by the upper bound $U = 3.13$. Consequently, as long as the statistical test provides an indication that the aforementioned stochastic variable is in the interval $[L, U]$ the functioning of the power unit can be concluded to be free of faults. On the other side, when the bounds of the previously given confidence interval are exceeded it can be concluded that the power unit has been subjected to a fault or cyber-attack. Finally, by performing the statistical test in subspaces of the power unit's state-space model, where each subspace is associated with different components, one can also achieve fault isolation. This signifies that the specific component that has caused the malfunctioning of the power unit can be identified.

4.1.7 SIMULATION TESTS

The performance of the proposed faults and cyber-attacks detection method has been confirmed through simulation experiments. Several faults and cyber-attacks have been considered, affecting both components of the power unit that were found in the interior of the control loop, as well as sensors and software or data that were found out of the control loop. Considering $n = 3$ measurable outputs for the power generation unit, to conclude the normal functioning of the power unit, the previously analyzed statistical test should take a value that is very close to the mean value of χ^2 distribution, that is 3.

The simulation of the dynamics, control and state-estimation for the model of the gas-turbine and synchronous generator has been carried out in discrete time and at a sampling period equal to $T_s = 0.01\text{sec}$ (or at at sampling frequency of $f_s = 0.1kHz$). By applying differential flatness theory the continuous-time state-space model of the gas-turbine and synchronous generator power unit has been first written into an input-output linearized form (canonical Brunovsky form), which has been next discretized with the use of Tustin's transformation.

To perform the faults detection and isolation tests with the previously analyzed χ^2 statistical criterion one can define the output measurements vector $y_m = [x_1, x_4, x_8]$ where $x_1 = \theta$ (turn angle of the generator's rotor), $x_4 = p_{cc}$ (pressure at the combustion chamber), and $x_8 = T_{exm}$ (measured temperature at the exhaust of the turbine). Thus the dimension of the outputs measurements vector is $m = 3$. Considering that the number of output vector samples is $M = 2000$ and using a 98% confidence interval for the χ^2 distribution the fault thresholds can be as $L = 2.87$ and $U = 3.13$. In an equivalent manner when the statistical test is applied exclusively to the gas turbine the fault thresholds are defined as $L = 1.89$ and $U = 2.11$. Moreover, when the statistical test is applied only to the synchronous generator the fault thresholds are defined as $L = 0.90$ and $U = 1.10$.

In case of fault detection with χ^2 tests, the Derivative-free nonlinear Kalman Filter has been used as a state estimator and the statistical χ^2 test has been applied to the power system that comprises both the gas turbine and the synchronous generator. In particular, the following cases have been examined:

- case (0): functioning of the gas-turbine and synchronous generator power unit in the fault-free condition. The associated results are depicted in Fig. 4.5 and Fig. 4.6.
- case (1a): functioning of the power generation unit under additive input disturbance at the generator. The associated results are depicted in Fig. 4.7 to Fig. 4.10.
- case (1b) functioning of the power generation unit under additive input disturbance at the turbine. The associated results are depicted in Fig. 4.11 to Fig. 4.14.
- case (2a): functioning of the power generation unit under parametric change in the control inputs gain matrix of the generator. The associated results are depicted in Fig. 4.15.

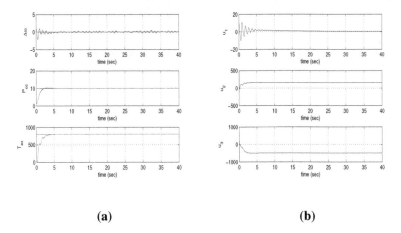

(a)   **(b)**

Figure 4.5   Test case 0: (a) variation of the outputs of the gas-turbine power unit y_i, $i = 1, \cdots, 3$ when neither a failure nor a cyberattack had taken place (b) variation of the control inputs v_i, $i = 1, \cdots, 3$ when neither a failure nor a cyberattack had taken place.

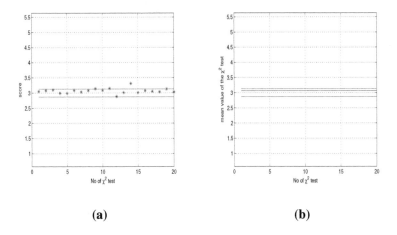

(a)   **(b)**

Figure 4.6   Test case 0: (a) values of successive χ^2 tests performed for the gas-turbine power unit when neither a failure nor a cyberattack had taken place (b) mean value of the individual χ^2 tests in the fault-free case.

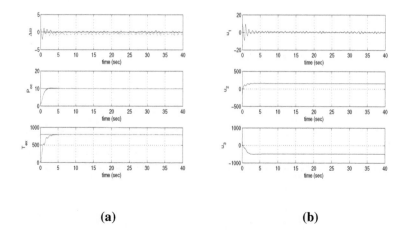

(a)   (b)

Figure 4.7 Test case 1a: (a) variation of the outputs of the gas-turbine power unit y_i, $i = 1, \cdots, 3$ under fault (additive disturbance) at the control input of the generator (b) variation of the control inputs v_i, $i = 1, \cdots, 3$ under fault (additive disturbance) at the control input of the generator

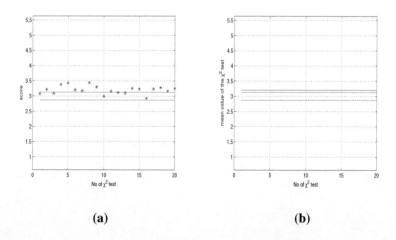

(a)   (b)

Figure 4.8 Test case 1a: (a) values of successive χ^2 tests performed for the entire gas-turbine power unit under fault (additive disturbance) at the control input of the generator (b) mean value of the individual χ^2 tests under fault (additive disturbance) at the control input of the generator

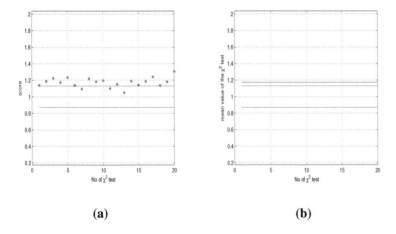

(a)　　　　　　　　　　　　(b)

Figure 4.9 Test case 1a: (a) values of successive χ^2 tests performed at the generator under fault (additive disturbance) at the control input of the generator (b) mean value of the individual χ^2 tests performed at the generator under fault (additive disturbance) at the control input of the generator

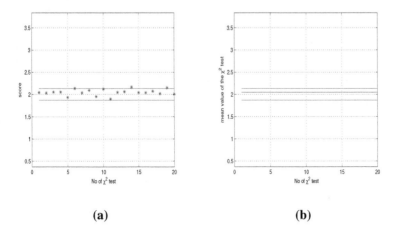

(a)　　　　　　　　　　　　(b)

Figure 4.10 Test case 1a: (a) values of successive χ^2 tests performed at the turbine under fault (additive disturbance) at the control input of the generator (b) mean value of the individual χ^2 tests performed at the turbine under fault (additive disturbance) at the control input of the generator

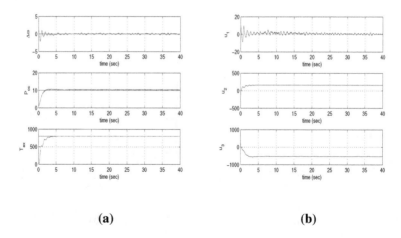

(a)   **(b)**

Figure 4.11 Test case 1ab: (a) variation of the outputs of the gas-turbine power unit y_i, $i = 1, \cdots, 3$ under fault (additive disturbance) at the control input of the turbine (b) variation of the control inputs v_i, $i = 1, \cdots, 3$ under fault (additive disturbance) at the control input of the turbine

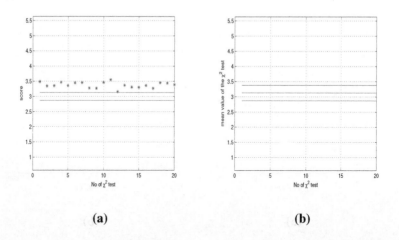

(a)   **(b)**

Figure 4.12 Test case 1b: (a) values of successive χ^2 tests performed for the entire gas-turbine power unit under fault (additive disturbance) at the control input of the turbine (b) mean value of the individual χ^2 tests under fault (additive disturbance) at the control input of the turbine

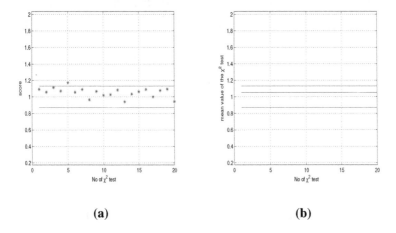

(a) **(b)**

Figure 4.13 Test case 1b: (a) values of successive χ^2 tests performed at the generator under fault (additive disturbance) at the control input of the turbine (b) mean value of the individual χ^2 tests performed at the generator under fault (additive disturbance) at the control input of the turbine

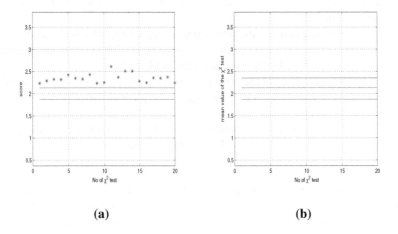

(a) **(b)**

Figure 4.14 Test case 1b: (a) values of successive χ^2 tests performed at the turbine under fault (additive disturbance) at the control input of the turbine (b) mean value of the individual χ^2 tests performed at the turbine under fault (additive disturbance) at the control input of the turbine

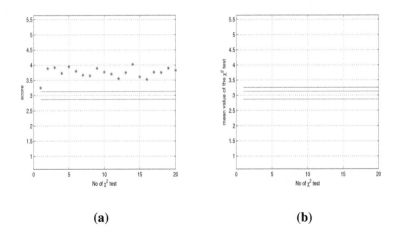

(a)   (b)

Figure 4.15 Test case 2a: (a) values of successive χ^2 tests performed for the entire gas-turbine power unit under parametric change at the control inputs gain matrix of the generator (b) mean value of the individual χ^2 tests under parametric change at the control inputs gain matrix of the generator.

- case (2b): functioning of the power generation unit under parametric change in the control inputs gain matrix of the turbine. The associated results are depicted in Fig. 4.16.
- case (3): functioning of the power generation units under faults at the sensors measuring the output of the gas-turbine and of the synchronous generator. The associated results are depicted in Fig. 4.17.
- case (4): functioning of the power generation unit under perturbations at the outputs' estimates which are provided by the Kalman Filter. The associated results are depicted in Fig. 4.18.
- case (5): functioning of the power generation unit under perturbations at the inputs applied to the Kalman Filter. The associated results are depicted in Fig. 4.19.

It has been confirmed that in all aforementioned cases the proposed condition monitoring method was capable of detecting incipient faults appearing in components of the power unit as well as perturbations affecting the Kalman Filter. The latter could be classified as cyber-attacks.

In all model-based fault diagnosis approaches it is considered that the model(s) which represent the fault-free functioning of the monitored system are known. If for some reason such models get invalidated they have to be substituted by an updated and corrected version of them. The change of an operating point is not related with the occurrence of a fault or cyber-attack in the electric power unit. The statistical test is repeated several times and the mean value of it provides the indication about the

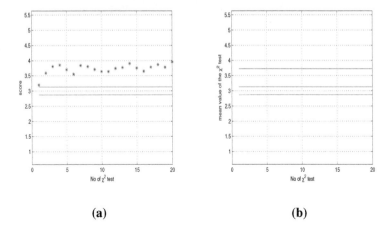

(a)   (b)

Figure 4.16 Test case 2b: (a) values of successive χ^2 tests performed for the entire gas-turbine power unit under parametric change at the control inputs gain matrix of the turbine (b) mean value of the individual χ^2 tests under parametric change at the control inputs gain matrix of the turbine.

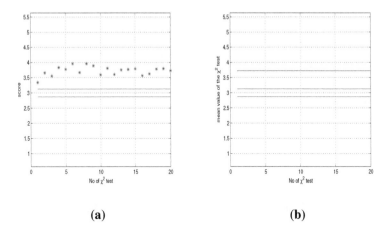

(a)   (b)

Figure 4.17 Test case 3: (a) values of successive χ^2 tests performed for the entire gas-turbine power unit under fault at the sensors measuring the system's outputs (b) mean value of the individual χ^2 tests under fault at the sensors measuring the system's outputs

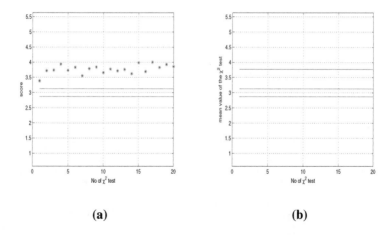

(a)　　　　　　　　　　　　　　　**(b)**

Figure 4.18　Test case 4: (a) values of successive χ^2 tests performed for the entire gas-turbine power unit under perturbation at the estimated outputs of the Kalman Filter (b) mean value of the individual χ^2 tests under perturbation at the estimated outputs of the Kalman Filter

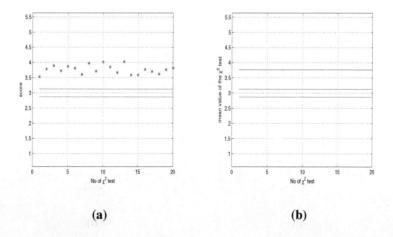

(a)　　　　　　　　　　　　　　　**(b)**

Figure 4.19　Test case 5: (a) values of successive χ^2 tests performed for the entire gas-turbine power unit under perturbation at the inputs of the Kalman Filter (b) mean value of the individual χ^2 tests under perturbation at the inputs of the Kalman Filter

faulty condition. This signifies that a change of operating point throughout several fault diagnosis tests cannot be misinterpreted as a failure.

A comparison between the Derivative-free nonlinear Kalman Filter and other nonlinear filtering approaches, shows that such methods provide suboptimal estimations while also being of questionable convergence. Actually, in the Extended Kalman Filter there is cumulative numerical error that may result into inaccurate state estimation of the monitored system. Besides, in the federated Kalman filter there is no proof that the fusion of estimates from multiple Kalman Filters finally improves the accuracy of the provided state estimation. On the other hand, the Derivative-free nonlinear Kalman Filtering approach is conditionally optimal, which signifies that it can achieve a minimum variance state estimation for the monitored nonlinear systems. This holds when the monitored outputs of the system are also its flat outputs. Such a condition is satisfied for the case of the gas-turbine and synchronous generator power unit. Besides, this filtering approach is convergent provided that the monitored system is observable. Such a property holds for the input-output linearized form of all differentially flat systems, and consequently it also holds for gas-turbine and synchronous generator power unit [195].

The distinctive features of the proposed fault diagnosis method which make it be advantageous when compared against other condition monitoring approaches for the gas-turbine and synchronous generator power unit are as follows: (i) solution of the nonlinear estimation problem for the power unit using an optimal nonlinear filter, which is the Derivative-free nonlinear Kalman Filter. By assuring minimum variance estimation for the state variables of the power unit the related fault diagnosis test is also more accurate and (ii) solution of the fault threshold definition problem in an objective manner, by using the outputs' estimation errors (residuals) to define a new stochastic variable and by exploiting the χ^2 distribution properties of this variable. The above features of the new fault diagnosis method allow for incipient fault detection and for minimization of the false alarms rate in the condition monitoring of the gas-turbine power unit.

4.2 FAULT DIAGNOSIS FOR THE STEAM-TURBINE AND SG POWER UNIT

4.2.1 OUTLINE

At a second stage the chapter proposes a solution for the fault diagnosis problem of a steam-turbine power unit consisting of a synchronous generator that is actuated by a steam turbine. It is widely known that power generation units exhibit a highly nonlinear and complex dynamics [248], [191], [194], [193], [218]. The chapter develops a new method for statistical condition monitoring of thermal power units consisting of steam turbines and synchronous power generators. The method can improve the security levels of such power units against faults of their electromechanical components, and can identify malicious human intrusions and cyberattacks to the power system. In the recent years there have been several results toward condition monitoring and fault diagnosis of the power grid [81], [95], [211], [54], [53]. In a significant number out of these results the Kalman Filter appears as a main approach toward residuals' generation [216], [124], [223], [224], [225]. A major limitation for solving the fault diagnosis problem in thermal power units remains that the standard Kalman Filter is a linear estimation method whereas such power systems exhibit a strongly nonlinear dynamics [59], [184],109,[138], [160], [23]. Apart from fault diagnosis, it has become apparent that there is need to develop condition monitoring methods capable of detecting cyberattacks and human intrusion in the aforementioned type of power generation units [94], [178], [145], [146].

In the present chapter, it is proven first that the dynamical system which comprises the steam turbine and the synchronous generator is a differentially flat one. This property confirms that the system can be transformed into an input-output linearized form after successive derivations of its flat outputs. For the linearized equivalent model of the power system one can solve both the control and the state estimation problems. In particular to obtain estimates of the state vector of the power system a differential flatness-based implementation of the Kalman Filter (Derivative-free nonlinear Kalman Filter) is proposed [126], [71], [230], [245], [156]. This consists of the standard Kalman Filter recursion applied to the linearized model of the system and of an inverse transformation that allows for obtaining estimates of the state variables of the initial nonlinear model of the power system. Moreover, by extending the state vector of the system after including as additional state variables the additive input disturbances and their time derivatives, it is shown that the Kalman Filter can be redesigned as a disturbance observer [195], [198], [75], [266].

Next, the Kalman Filter is used as a software sensor which provides solution to the condition monitoring problem of the power system. The flat outputs of the power system are compared against the outputs of the Kalman Filter and their differences form the residuals' sequence [188], [18], [277], [189]. The stochastic variable which is equal to the square of the residuals' vector weighted by the inverse of the associated covariance matrix is shown to follow the χ^2 distribution. By exploiting the statistical properties of the χ^2 distribution, and after using the confidence intervals concept, one can define fault thresholds beyond which the abnormal functioning of the system is concluded. Thus one arrives at a generic statistical test which demonstrates reliably whether the electromechanical components of the power unit have been subjected to a fault or whether the power system has undergone a cyberattack. Additionally, by performing the statistical test to the individual components of the power unit one can perform fault isolation and can distinguish if the fault has taken place at the steam turbine or at the synchronous generator.

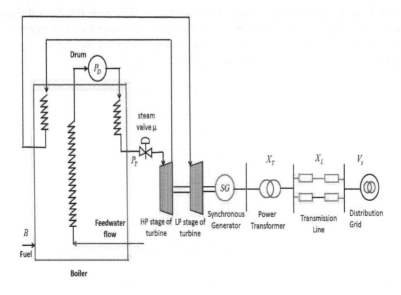

Figure 4.20   Steam-turbine and synchronous-generator power unit

4.2.2   DYNAMIC MODEL OF STEAM TURBINE AND SYNCHRONOUS GENERATOR

4.2.2.1   Power system's dynamics

The power generation unit, depicted in Fig. 4.20, comprises: (i) the electrical part which is the synchronous power generator and (ii) the thermal power part which is the boiler supplying steam to a turbine [248], [204]. The dynamics of the synchronous generator, being based on the single-axis model. is outlined once gain below:

Dynamics of the synchronous generator:

$$\dot{\delta} = \omega$$
$$\dot{\omega} = -\frac{D}{2J}(\omega - \omega_0) + \frac{\omega_0}{2J}(P_m - P_e) \tag{4.79}$$

where δ is the turn angle of the generator's rotor, ω is the rotation speed of the rotor with respect to synchronous reference, ω_0 is the synchronous speed of the generator, J is the moment of inertia of the rotor, P_e is the active power of the generator, P_m is the mechanical input torque to the generator which is associated with the mechanical input power, D is the damping constant of the generator and T_e is the electrical torque which is associated to the generated active power. Moreover, the following variables are defined: $\Delta\delta = \delta - \delta_0$ and $\Delta\omega = \omega - \omega_0$ with ω_0 denoting the synchronous speed. The generator's electrical dynamics is described as follows [204]:

$$\dot{E}'_q = \frac{1}{T_{d_o}}(E_f - E_q) \tag{4.80}$$

where E_q' is the quadrature-axis transient voltage of the generator, E_q is the quadrature axis voltage of the generator, T_{do} is the direct axis open-circuit transient time constant of the generator and E_f is the equivalent voltage in the excitation coil. The algebraic equations of the synchronous generator are given by

$$
\begin{aligned}
E_q &= \tfrac{x_{d\Sigma}}{x_{d\Sigma}'} E_q' - (x_d - x_d') \tfrac{V_s}{x_{d\Sigma}'} cos(\Delta\delta) \\
I_q &= \tfrac{V_s}{x_{d\Sigma}'} sin(\Delta\delta) \\
I_d &= \tfrac{E_q'}{x_{d\Sigma}'} - \tfrac{V_s}{x_{d\Sigma}'} cos(\Delta\delta) \\
P_e &= \tfrac{V_s E_q'}{x_{d\Sigma}'} sin(\Delta\delta) \\
Q_e &= \tfrac{V_s E_q'}{x_{d\Sigma}'} cos(\Delta\delta) - \tfrac{V_s^2}{x_{d\Sigma}'} \\
V_t &= \sqrt{(E_q' - X_d' I_d)^2 + (X_d' I_q)^2}
\end{aligned}
\tag{4.81}
$$

where $x_{d\Sigma} = x_d + x_T + x_L$, $x_{d\Sigma}' = x_d' + x_T + x_L$, x_d is the direct-axis synchronous reactance, x_T is the reactance of the transformer, x_d' is the direct axis transient reactance, x_L is the reactance of the transmission line, I_d and I_q are direct and quadrature axis currents of the generator, V_s is the infinite bus voltage, Q_e is the generator reactive power delivered to the infinite bus, and V_t is the terminal voltage of the generator. Substituting the electrical equations of the SG given in Eq. (4.81) into the equations of the electrical and mechanical dynamics of the rotor given in Eq. (4.79) and Eq. (4.80) respectively, the complete model of the Single Machine Infinite Bus model is obtained

$$
\begin{aligned}
\dot{\delta} &= \omega - \omega_0 \\
\dot{\omega} &= -\tfrac{D}{2J}(\omega - \omega_0) + \omega_0 \tfrac{P_m}{2J} - \omega_0 \tfrac{1}{2J} \tfrac{V_s E_q'}{x_{d\Sigma}'} sin(\Delta\delta) \\
\dot{E}_q' &= -\tfrac{1}{T_d'} E_q' + \tfrac{1}{T_{do}} \tfrac{x_d - x_d'}{x_{d\Sigma}'} V_s cos(\Delta\delta) + \tfrac{1}{T_{do}} E_f
\end{aligned}
\tag{4.82}
$$

where $T_d' = \tfrac{x_{d\Sigma}'}{x_{d\Sigma}} T_{do}$ is the time constant of the field winding, and E_f is the excitation voltage.

Dynamics of the steam turbine:

$$
\dot{P}_m = \frac{1}{T_h}[-P_m + P_T \mu]
\tag{4.83}
$$

$$
\dot{P}_T = \frac{1}{C_{SH}}[k\sqrt{P_D - P_T} - P_t \mu]
\tag{4.84}
$$

$$
\dot{P}_D = \frac{1}{C_D}[D_Q - k\sqrt{P_D - P_T}]
\tag{4.85}
$$

$$\dot{D}_Q = \frac{1}{T_B}[-D_Q + B] \tag{4.86}$$

where P_m is the mechanical power provided to the generator by the steam turbine, P_T is the pressure before the turbine, P_D is the pressure in the drum and D_Q is the mass flow of steam in the boiler (steam efficiency). Other parameters of this model are defined as follows: B is the flow of the fuel to the boiler, C_D is the time constant due to drum capacity, C_{SH} is the time constant due to volume in superheater, k is the coefficient related to the flow of steam, T_B is the boiler's time constant, T_h is the time constant due to volume in turbine and reheater;

The steam-turbine and synchronous-generator power unit is shown in Fig. 4.20:

Next,ones defines $X = [x_1, x_2, x_3, x_4, x_5, x_6, x_7]^T = [\Delta\delta, \Delta\omega, E_q', P_m, P_T, P_D, D_Q]^T$ as the state vector of the system, while the control inputs vector of the system is set to be $U = [u_1, u_2]^T = [E_f, B]^T$. This results into the following state-space description of the system

$$\dot{x}_1 = x_2 \tag{4.87}$$

$$\dot{x}_2 = -\frac{D}{2J}(x_2 - \omega_0) + \omega_0\frac{x_4}{2J} - \omega_0\frac{1}{2J}\frac{V_s x_3}{x_{d\Sigma}'}\sin(x_1) \tag{4.88}$$

$$\dot{x}_3 = -\frac{1}{T_d'}x_3 + \frac{1}{T_{do}'}\frac{x_d - x_d'}{x_{d\Sigma}'}V_s\cos(x_1) + \frac{1}{T_{do}'}u_1 \tag{4.89}$$

$$\dot{x}_4 = \frac{1}{T_h}[-x_4 + x_5\mu] \tag{4.90}$$

$$\dot{x}_5 = \frac{1}{C_{SH}}[k\sqrt{x_6 - x_5} - x_5\mu] \tag{4.91}$$

$$\dot{x}_6 = \frac{1}{C_D}[x_7 - k\sqrt{x_6 - x_5}] \tag{4.92}$$

$$\dot{x}_7 = \frac{1}{T_B}[-x_7 + u_2] \tag{4.93}$$

As a result of the above the state-space model of the system is written in the following matrix form:

$$
\begin{pmatrix} \dot{x}_1 \\ \dot{x}_2 \\ \dot{x}_3 \\ \dot{x}_4 \\ \dot{x}_5 \\ \dot{x}_6 \\ \dot{x}_7 \end{pmatrix} =
\begin{pmatrix} x_2 \\ -\frac{D}{2J}(x_2 - \omega_0) + \omega_0\frac{x_4}{2J} - \omega_0\frac{1}{2J}\frac{V_s x_3}{x_{d\Sigma}'}\sin(x_1) \\ \dot{x}_3 = -\frac{1}{T_d'}x_3 + \frac{1}{T_{do}'}\frac{x_d - x_d'}{x_{d\Sigma}'}V_s\cos(x_1) \\ \frac{1}{T_h}(-x_4) + \frac{x_5\mu}{T_b} \\ \frac{1}{C_{SH}}[k\sqrt{x_6 - x_5}] - \frac{x_5\mu}{C_{SH}} \\ \frac{1}{C_D}[x_7 - k\sqrt{x_6 - x_5}] \\ -\frac{1}{T_B}x_7 \end{pmatrix} +
\begin{pmatrix} 0 & 0 \\ 0 & 0 \\ \frac{1}{T_{do}} & 0 \\ 0 & 0 \\ 0 & 0 \\ 0 & 0 \\ 0 & \frac{1}{T_B} \end{pmatrix}
\begin{pmatrix} u_1 \\ u_2 \end{pmatrix} \tag{4.94}
$$

The state-space description of the power system can be also written in the concise matrix form:

$$\dot{x} = f(x) + G(x)u \tag{4.95}$$

where $x \in R^{7\times1}$, $f(x) \in R^{7\times1}$, $G(x) \in R^{7\times2}$ and $u \in R^{2\times1}$, with

$$f(x) = \begin{pmatrix} x_2 \\ -\frac{D}{2J}(x_2 - \omega_0) + \omega_0 \frac{x_4}{2J} - \omega_0 \frac{1}{2J} \frac{V_s x_3}{x'_{d\Sigma}} sin(x_1) \\ \dot{x}_3 = -\frac{1}{T'_d} x_3 + \frac{1}{T_{do}} \frac{x_d - x'_d}{x'_{d\Sigma}} V_s cos(x_1) \\ \frac{1}{T_h}(-x_4) + \frac{x_5 \mu}{T_b} \\ \frac{1}{C_{SH}}[k\sqrt{x_6 - x_5}] - \frac{x_5 \mu}{C_{SH}} \\ \frac{1}{C_D}[x_7 - k\sqrt{x_6 - x_5}] \\ -\frac{1}{T_B} x_7 \end{pmatrix} \quad G(x) = \begin{pmatrix} 0 & 0 \\ 0 & 0 \\ \frac{1}{T_{do}} & 0 \\ 0 & 0 \\ 0 & 0 \\ 0 & 0 \\ 0 & \frac{1}{T_B} \end{pmatrix}$$

$$\tag{4.96}$$

4.2.2.2   Faults and cyberattacks affecting the power system

Indicative faults and cyberattacks or human intrusions that may affect the power generation unit, which comprises the steam turbine and the synchronous generator, are shown in Fig. 4.21. The Kalman Filter's outputs are compared at each sampling instance against the outputs measured from the power generation unit thus providing the residuals' sequence which contains information about the occurrence of faulta. These faults are: (i) fault at the control input gains of the steam turbine $\Delta i g_1$, (ii) fault at the control input gains of the generator $\Delta i g_2$, (iii) fault (additive disturbance) at the control input at the turbine Δu_1, (iv) fault (additive disturbance) at the control input at the generator Δu_2, (v) fault at the steam turbine's control input sensors $\Delta i s_1$, (vi) fault at the generator's control input sensors $\Delta i s_2$, (vii) fault at the steam turbine's output measurement sensors $\Delta o s_1$ and (viii) fault at the generator's output measurement sensors $\Delta o s_2$.

4.2.3   DIFFERENTIAL FLATNESS PROPERTIES OF THE POWER UNIT'S MODEL

4.2.3.1   Proof of differential flatness for the power generation unit

The dynamic model of the steam-turbine and of the synchronous generator that has been defined in Eq. (4.87) to Eq. (4.93) is a differentially flat one. This property can be used for solving the control (synchronization) and state estimation problem of the power unit. The flat outputs vector of the power system is taken to be $[x_1, x_4]^T = [\Delta\delta, P_m]^T$ and comprises as elements the turn angle of the generator $\Delta\delta$ and the mechanical power that is provided to the generator by the turbine.

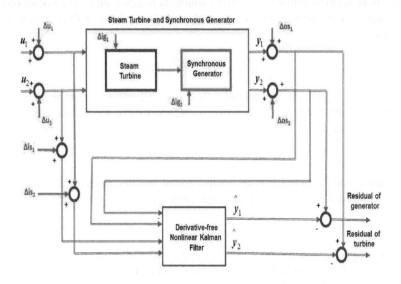

Figure 4.21 Faults and cyberattacks affecting the power generation unit that comprises the steam turbine and the synchronous generator

From Eq. (4.87) and by solving with respect to x_2 one has

$$x_2 = \dot{x}_1 \tag{4.97}$$

From Eq. (4.89) and by solving with respect to x_3 one has

$$x_3 = \frac{\dot{x}_2 + \frac{D}{2J}x_2 - \frac{\omega}{2J}x_4}{\frac{-\omega_0}{2J}\frac{V_s}{x'_{d\Sigma}}sin(x_1)} \tag{4.98}$$

From Eq. (4.97) and Eq. (4.98) one has that state variables x_2 and x_3 are differential functions of the flat output. Moreover, from Eq. (4.89) and after solving with respect to u_1 one has

$$u_1 = T_{do}[\dot{x}_3 + \frac{1}{T'_d}x_3 - \frac{1}{T_{do}}\frac{x_d - x'_d}{x'_{d\Sigma}}V_s cos(x_1) + \frac{1}{T_{do}}u_1] \tag{4.99}$$

Consequently, the control input u_1 is also a differential function of the flat output. Next, after solving Eq. (4.90) with respect to state variable x_5 one gets

$$x_5 = \frac{T_h\dot{x}_4 + x_4}{\mu} \tag{4.100}$$

The above relation signifies that state variable x_5 is also a differential function of the flat output of the system. From Eq. (4.91) and after solving with respect to x_6, one obtains

$$x_6 = x_5 + (\frac{C_{SH}\dot{x}_5 + x_5\mu}{k})^2 \qquad (4.101)$$

From Eq. (4.101) it can be concluded that state variable x_6 is a differential function of the flat output. Moreover, from Eq. (4.92) and by solving with respect to state variable x_7 one has

$$x_7 = C_d\dot{x}_6 + k\sqrt{x_6 - x_5} \qquad (4.102)$$

Therefore, state variable x_7 is also a differential function of the system's flat output. Finally, by solving Eq. (4.93) with respect to the control input u_2 one gets

$$u_2 = T_B\dot{x}_7 + x_7 \qquad (4.103)$$

which means that control input u_2 can be expressed as a function of the flat output and its derivatives. In conclusion all state variables and the control inputs of the model of the steam-turbine and synchronous generator are differential functions of the flat output vector and as a result of this the considered power system is a differentially flat one.

After defining reference setpoints for the elements of the flat outputs vector x_1^d and x_4^d and using Eq. (4.97) to Eq. (4.103) one can compute the reference setpoints for the rest of the state variables of the power system, that is x_2^d, x_3^d, x_5^d, x_6^d and x_7^d as differential functions of x_1^d and x_4^d. This provides a systematic method for selecting the setpoints for the nonlinear optimal control problem of the steam-turbine and of the synchronous-generator system.

4.2.3.2 Transformation of the power unit to an input-output linearized form

According to the property of the Lie-Backlünd equivalence which holds for differentially flat systems, a dynamical system that is proven to be differentially flat can be also transformed into an input-output linearized form. This also holds for the dynamic model of the turbine and synchronous generator. The flat output of the power generation system has been defined to be $Y = [x_1, x_4]$. By differentiating the first state-space equation given in Eq. (4.87), one obtains

$$\ddot{x}_1 = \dot{x}_2 \Rightarrow$$
$$\ddot{x}_1 = -\frac{D}{2J}(x_2 - \omega_0) + \omega_0\frac{x_4}{2J} - \omega_0\frac{1}{2J}\frac{V_s x_3}{x_{d\Sigma}}\sin(x_1) \qquad (4.104)$$

By differentiating x_1 once more in time

$$x_1^{(3)} = -\frac{D}{2J}(\dot{x}_2 - \omega_0) + \omega_0\frac{\dot{x}_4}{2J} -$$
$$- \omega_0\frac{1}{2J}\frac{V_s}{x_{d\Sigma}}[\dot{x}_3\sin(x_1) + x_3\cos(x_1)] \qquad (4.105)$$

By substituting the time-derivatives of the state variables in the aforementioned relations one gets:

$$x_1^{(3)} = -\frac{D}{2J}[-\frac{D}{2J}(x_2 - \omega_0) + \omega_0\frac{x_4}{2J} - \omega_0\frac{1}{2J}\frac{V_s x_3}{x'_{d\Sigma}}sin(x_1)]) + \omega_0\frac{1}{2J}\frac{1}{T_h}[-x_4 + x_5\mu] +$$
$$+\frac{\omega_0}{2J}\frac{V_s}{x'_{d\Sigma}}x_3 cos(x_1) - \frac{\omega_0}{2J}\frac{V_s}{x'_{d\Sigma}}sin(x_1)\cdot\cdot[-\frac{1}{T_d}x_3 + \frac{1}{T_{do}}\frac{x_d - x'_d}{x'_{d\Sigma}}V_s cos(x_1)] -$$
$$-\frac{\omega_0}{2J}\frac{V_s}{x'_{d\Sigma}}sin(x_1)\frac{1}{T_{do}}u_1$$

$$(4.106)$$

This is a relation in the form

$$x_1^{(3)} = f_1(x) + g_1(x)u_1 \tag{4.107}$$

Next, by differentiating Eq. (4.90) in time, one obtains

$$\dddot{x}_4 = -\frac{1}{T_h}^2(\dot{x}_4 + \dot{x}_5\mu \rightarrow$$
$$\ddot{x}_4 = \frac{1}{T_h}\{-\frac{1}{T_h}[-x_4 + x_5\mu] + \frac{1}{C_{SH}}[k\sqrt{x_6 - x_5} - \mu x_5]\} \Rightarrow \tag{4.108}$$
$$\ddot{x}_4 = -\frac{1}{T_h^2}(-\dot{x}_4 + \dot{x}_5\mu) + \frac{1}{T_h}\frac{1}{C_{SH}}k(x_6 - x_5)^{\frac{1}{2}} - \frac{1}{T_h}\frac{1}{C_{SH}}\mu x_5$$

By differentiating x_4 once more in time one has

$$x_4^{(3)} = -\frac{1}{T_h^2}(-\dot{x}_4 + \dot{x}_5\mu) +$$
$$+\frac{1}{T_h}\frac{1}{C_{SH}}\frac{k}{2}(x_6 - x_5)^{-\frac{1}{2}}(\dot{x}_6 - \dot{x}_5) - \frac{1}{T_h}\frac{1}{C_{SH}}\mu\dot{x}_5 \Rightarrow$$
$$x_4^{(3)} = -\frac{1}{T_h^2}\{-\frac{1}{T_h}[-x_4 + x_5\mu] + \frac{1}{C_{SH}}[k\sqrt{x_6 - x_5} - \mu x_5]\} + \tag{4.109}$$
$$+\frac{1}{T_f}\frac{1}{C_{SH}}\frac{k}{2}(x_6 - x_5)^{-\frac{1}{2}}\{\frac{1}{C_d}[x_7 - k\sqrt{x_6 - x_5}] +$$
$$\frac{1}{C_{SH}}[k\sqrt{x_6 - x_5}]\} - \frac{1}{T_h}\frac{1}{C_{SH}}\mu\frac{1}{C_{SH}}[k\sqrt{x_6 - x_5} - \mu x_5]$$

By continuing the differentiations of x_4, one obtains

$$x_4^{(4)} = -\frac{1}{T_h^2}\{-\frac{1}{T_h}\left[x_4 + x_5\mu\right] + \frac{1}{C_{SH}}[\frac{k}{2}(x_6 - x_5)^{-\frac{1}{2}}(\dot{x}_6 - \dot{x}_5) - \mu\dot{x}_5]\mu) +\}$$
$$+\frac{1}{T_h}\frac{1}{C_{SH}} - \frac{k}{4}(x_6 - x_5)^{-\frac{3}{2}}(\dot{x}_6 - \dot{x}_5)$$
$$\{\frac{1}{C_D}[x_7 - k\sqrt{x_6 - x_5}] + \frac{1}{C_{SH}}[k\sqrt{x_6 - x_5} - \mu x_5]\}$$
$$\frac{1}{T_h}\frac{1}{C_{SH}} - \frac{k}{4}(x_6 - x_5)^{-\frac{1}{2}}\{\frac{1}{C_D}[\dot{x}_7 - \tag{4.110}$$
$$\frac{k}{2}(x_6 - x_5)^{-\frac{1}{2}}(\dot{x}_6) - \dot{x}_5] + \frac{1}{C_{SH}}$$
$$\left[\frac{k}{2}(x_6 - x_5)^{-\frac{1}{2}}(\dot{x}_6 - \dot{x}_5) - \mu\dot{x}_5\right]\} -$$
$$-\frac{1}{T_h}\frac{1}{C_{SH}^2}\mu\left[\frac{k}{2}(x_6 - x_5)^{-\frac{1}{2}}(\dot{x}_6 - \dot{x}_5) - \mu\dot{x}_5\right]$$

Next, by substituting in the above equation the description for $\dot{x}_7$ from Eq. (4.93) one arrives at:

$$
\begin{aligned}
x_4^{(4)} = -\frac{1}{T_h^2}\{-\frac{1}{T_h}[x_4+x_5\mu] + \frac{1}{C_{SH}}[\frac{k}{2}(x_6-x_5)^{-\frac{1}{2}}(\dot{x}_6-\dot{x}_5)-\mu\dot{x}_5]\mu)+\} \\
+\frac{1}{T_h}\frac{1}{C_{SH}}-\frac{k}{4}(x_6-x_5)^{-\frac{3}{2}}(\dot{x}_6-\dot{x}_5)\cdot \\
\{\frac{1}{C_D}[x_7-k\sqrt{x_6-x_5}] + \frac{1}{C_{SH}}[k\sqrt{x_6-x_5}-\mu x_5]\}\frac{1}{T_h}\frac{1}{C_{SH}}-\frac{k}{4}(x_6-x_5)^{-\frac{1}{2}}\cdot \\
\{\frac{1}{C_D}[\frac{1}{T_D}x_7-\frac{k}{2}(x_6-x_5)^{-\frac{1}{2}}(\dot{x}_6)- \\
\dot{x}_5]+\frac{1}{C_{SH}}[\frac{k}{2}(x_6-x_5)^{-\frac{1}{2}}(\dot{x}_6-\dot{x}_5)-\mu\dot{x}_5]\}- \\
-\frac{1}{T_h}\frac{1}{C_{SH}^2}\mu[\frac{k}{2}(x_6-x_5)^{-\frac{1}{2}}(\dot{x}_6-\dot{x}_5)- \\
\mu\dot{x}_5] +\frac{1}{T_h}\frac{1}{C_{SH}}-\frac{k}{4}(x_6-x_5)^{-\frac{1}{2}}\frac{1}{T_D}u_2
\end{aligned}
\tag{4.111}
$$

This is a relation of the form

$$
x_4^{(4)} = f_2(x)+g_2(x)u_2
\tag{4.112}
$$

In aggregate one arrives at an input-output linearized form for the power generation unit that comprises the steam-turbine and the synchronous generator

$$
\begin{pmatrix} x_1^{(3)} \\ x_4^{(4)} \end{pmatrix} = \begin{pmatrix} f_1(x) \\ f_2(x) \end{pmatrix} + \begin{pmatrix} g_1(x) & 0 \\ 0 & g_2(x) \end{pmatrix} \begin{pmatrix} u_1 \\ u_2 \end{pmatrix}
\tag{4.113}
$$

In this manner a complete input-output decoupling of the power system is achieved. Moreover, by defining the new control inputs $v_1 = f_1(x)+g_1(x)u_1$ and $v_2 = f_2(x)+g_2(x)u_2$, one obtains

$$
x_1^{(3)} = v_1 \quad x_4^{(4)} = v_2
\tag{4.114}
$$

Next, the following feedback control law is considered.

$$
u_1 = \frac{1}{g_1(x)}[x_{1,d}^{(3)} - f_1(x) - k_1^1(\ddot{x}_1 - \ddot{x}_{1,d}) - k_2^1(\dot{x}_1 - \dot{x}_1^d) - k_3^1(x_1 - x_{1,d})]
$$

$$
\begin{aligned}
u_2 = \frac{1}{g_2(x)}[x_{1,d}^{(4)} - f_2(x) - k_1^2(x_4^{(3)} - x_{4,d}^{(3)}) - \\
-k_2^2(\ddot{x}_4 - \ddot{x}_{4,d}) - k_3^2(\dot{x}_4 - \dot{x}_4^d) - k_3^2(x_4 - x_{4,d})]
\end{aligned}
\tag{4.115}
$$

By applying the previous control law and by defining the tracking error variables as $e_1 = x_1 - x_{1,d}$ and $e_4 = x_4 - x_{4,d}$ one has

$$
\begin{aligned}
e_1^{(3)} + k_1^1\ddot{e}_1 + k_2^1\dot{e}_1 + k_3^1 e_1 = 0 \\
e_1^{(4)} + k_1^2 e_4^{(3)} + k_2^2\ddot{e}_4 + k_3^2\dot{e}_4 + k_4^2 e_4 = 0
\end{aligned}
\tag{4.116}
$$

Thus, by suitably selecting the feedback control gains $[k_1^1, k_2^1, k_3^1]$ and $[k_1^2, k_2^2, k_3^2, k_4^2]$ so as the characteristic polynomials associated with the previous two equations to be Hurwitz stable one assures that

$$
\begin{aligned}
lim_{t\to\infty} e_1 &= 0 \Rightarrow lim_{t\to\infty} x_1 = x_{1,d} \\
lim_{t\to\infty} e_4 &= 0 \Rightarrow lim_{t\to\infty} x_4 = x_{4,d}
\end{aligned}
\tag{4.117}
$$

The elimination of the tracking error for the flat outputs of the power system assures also the elimination of the tracking error for all state variables of the model and the synchronization of the power generator with the grid's frequency.

4.2.4 DIFFERENTIAL FLATNESS THEORY-BASED IMPLEMENTATION OF THE KALMAN FILTER

4.2.4.1 Design of a Kalman Filter-based state estimator for the power unit

Using that the flat outputs vector of the system is $[y_1, y_2] = [x_1, x_4]$, then in the input-output linearized form the power unit's dynamic model is given by

$$
\begin{aligned}
y_1^{(3)} &= v_1 \\
y_1^{(4)} &= v_2
\end{aligned}
\tag{4.118}
$$

The state vector of the power unit comprises the following state variables: $z_1 = y_1$, $z_2 = \dot{y}_1$, $z_3 = \ddot{y}_1$, $z_4 = y_2$, $z_5 = \dot{y}_2$, $z_6 = \ddot{y}_2$, $z_7 = y_2^{(3)}$. In this manner, and by defining the state vector $Z = [z_1, z_2, \cdots, z_7]^T$ one arrives at a state-space description for the power system, which is in the linear canonical (Brunovsky) form:

$$
\begin{pmatrix} \dot{z}_1 \\ \dot{z}_2 \\ \dot{z}_3 \\ \dot{z}_4 \\ \dot{z}_5 \\ \dot{z}_6 \\ \dot{z}_7 \end{pmatrix}
=
\begin{pmatrix}
0 & 1 & 0 & 0 & 0 & 0 & 0 \\
0 & 0 & 1 & 0 & 0 & 0 & 0 \\
0 & 0 & 0 & 0 & 0 & 0 & 0 \\
0 & 0 & 0 & 0 & 1 & 0 & 0 \\
0 & 0 & 0 & 0 & 0 & 1 & 0 \\
0 & 0 & 0 & 0 & 0 & 0 & 1 \\
0 & 0 & 0 & 0 & 0 & 0 & 0
\end{pmatrix}
\begin{pmatrix} z_1 \\ z_2 \\ z_3 \\ z_4 \\ z_5 \\ z_6 \\ z_7 \end{pmatrix}
+
\begin{pmatrix}
0 & 0 \\
0 & 0 \\
1 & 0 \\
0 & 0 \\
0 & 0 \\
0 & 0 \\
0 & 1
\end{pmatrix}
\begin{pmatrix} v_1 \\ v_2 \end{pmatrix}
\tag{4.119}
$$

$$
z^m = \begin{pmatrix} 1 & 0 & 0 & 0 & 0 & 0 & 0 \\ 0 & 0 & 0 & 1 & 0 & 0 & 0 \end{pmatrix} Z
\tag{4.120}
$$

The previous description of the system is written in the linear state-space form

$$
\begin{aligned}
\dot{z} &= Az + B\tilde{u} \\
z^m &= Cz
\end{aligned}
\tag{4.121}
$$

To perform simultaneous estimation of the non-measurable state-vector elements and of the disturbance inputs, the following disturbance observer is defined

$$
\begin{aligned}
\dot{\hat{z}} &= A_o \hat{z} + B_o v + K_f(z_m - \hat{z}_m) \\
z^m &= C_o z
\end{aligned}
\tag{4.122}
$$

where $v = [v_1, v_2]^T$, $A_o = A$, $C_o = C$ and

$$B_o = \begin{pmatrix} 1\,0\,0\,0\,0\,0\,0 \\ 0\,0\,0\,1\,0\,0\,0 \end{pmatrix}^T \tag{4.123}$$

The gain of the state observer is computed through the Kalman Filter's recursion. The application of Kalman Filtering on the linearized equivalent description of the power system is the so-called *Derivative-free nonlinear Kalman Filter*. Matrices, A_o, B_o and C_o are discretized using common discretization methods. This provides matrices A_d, B_d and C_d. Matrices Q and R denote the process and measurement noise covariance matrices. Matrix P is the state-vector error covariance matrix. The Kalman Filter for the state estimation problem comprises a *measurement update* and a *time update* stage.

measurement update:

$$K_f(k) = P^-(k)C_d^T[C_dP^-(k)C_d^T + R]^{-1}$$
$$\hat{x}(k) = \hat{x}^-(k) + K_f(k)[z_m - \hat{z}_m] \tag{4.124}$$
$$P(k) = P^-(k) - K_f(k)C_dP^-(k)$$

time update:

$$P^-(k+1) = A_dP(k)A_d^T + Q$$
$$\hat{x}^-(k+1) = A_d\hat{x}(k) + B_dv(k) \tag{4.125}$$

4.2.4.2   Design of a Kalman Filter-based disturbance estimator for the power unit

It is assumed that the power generation unit is subjected to additive input disturbances. Using that the flat outputs vector of the system is $[y_1, y_2] = [x_1, x_4]$, then in the input-output linearized form of the system the disturbances' effect is written as

$$y_1^{(3)} = v_1 + \tilde{d}_1$$
$$y_1^{(4)} = v_2 + \tilde{d}_2 \tag{4.126}$$

It is considered that the disturbances $\tilde{d}_1$ and $\tilde{d}_2$ are described by their time derivatives up to order 2, that is

$$\dddot{\tilde{d}}_1 = f_{d_1} \quad \dddot{\tilde{d}}_2 = f_{d_2} \tag{4.127}$$

Actually, every signal can be described by its derivatives up to order n and the associated initial conditions. However, since estimation of the signals $\tilde{d}_1$ and $\tilde{d}_2$ and of their derivatives is going to be performed with the use of Kalman Filtering, there is no prior constraint about knowing these initial conditions.

The state vector of the system is extended by considering as additional state variables the disturbance inputs and their derivatives. Thus, the new state variables are

$$z_1 = y_1,\ z_2 = \dot{y}_1,\ z_3 = \ddot{y}_1,\ z_4 = y_2,\ z_5 = \dot{y}_2,\ z_6 = \ddot{y}_2,\ z_7 = y_2^{(3)},\ z_8 = \tilde{d}_1,\ z_9 = \dot{\tilde{d}}_1,$$

$z_{10} = \dot{d}_2$ and $z_{11} = \ddot{d}_2$. In this manner, and by defining the extended state vector $Z = [z_1, z_2, \cdots, z_{11}]^T$ one arrives at a state-space description for the power system, which is in the linear canonical (Brunovsky) form:

$$
\begin{pmatrix} \dot{z}_1 \\ \dot{z}_2 \\ \dot{z}_3 \\ \dot{z}_4 \\ \dot{z}_5 \\ \dot{z}_6 \\ \dot{z}_7 \\ \dot{z}_8 \\ \dot{z}_9 \\ \dot{z}_{10} \\ \dot{z}_{11} \end{pmatrix} =
\begin{pmatrix}
0\,1\,0\,0\,0\,0\,0\,0\,0\,0\,0 \\
0\,0\,1\,0\,0\,0\,0\,0\,0\,0\,0 \\
0\,0\,0\,0\,0\,0\,1\,0\,0\,0 \\
0\,0\,0\,1\,0\,0\,0\,0\,0\,0 \\
0\,0\,0\,0\,0\,1\,0\,0\,0\,0 \\
0\,0\,0\,0\,0\,1\,0\,0\,0\,0 \\
0\,0\,0\,0\,0\,0\,0\,1\,0 \\
0\,0\,0\,0\,0\,0\,1\,0\,0 \\
0\,0\,0\,0\,0\,0\,0\,0\,0\,0 \\
0\,0\,0\,0\,0\,0\,0\,0\,0\,1 \\
0\,0\,0\,0\,0\,0\,0\,0\,0\,0
\end{pmatrix}
\begin{pmatrix} z_1 \\ z_2 \\ z_3 \\ z_4 \\ z_5 \\ z_6 \\ z_7 \\ z_8 \\ z_9 \\ z_{10} \\ z_{11} \end{pmatrix} +
\begin{pmatrix}
0\,0\,0\,0 \\
0\,0\,0\,0 \\
1\,0\,0\,0 \\
0\,0\,0\,0 \\
0\,0\,0\,0 \\
0\,0\,0\,0 \\
0\,1\,0\,0 \\
0\,0\,0\,0 \\
0\,0\,1\,0 \\
0\,0\,0\,0 \\
0\,0\,0\,1
\end{pmatrix}
\begin{pmatrix} v_1 \\ v_2 \\ \ddot{f}_{d_1} \\ \ddot{f}_{d_2} \end{pmatrix}
\tag{4.128}
$$

$$
z^m = \begin{pmatrix} 1\,0\,0\,0\,0\,0\,0\,0\,0\,0\,0 \\ 0\,0\,0\,1\,0\,0\,0\,0\,0\,0\,0 \end{pmatrix} Z
\tag{4.129}
$$

The previous description of the system is written in the linear state-space form

$$
\dot{z} = Az + B\tilde{u} \quad z^m = Cz
\tag{4.130}
$$

To perform simultaneous estimation of the non-measurable state-vector elements and of the disturbance inputs, the following disturbance observer is defined

$$
\dot{\hat{z}} = A_o \hat{z} + B_o v + K_f(z_m - \hat{z}_m)
$$
$$
z^m = C_o z
\tag{4.131}
$$

$v = [v_1, v_2]^T$, $A_o = A$, $C_o = C$ and

$$
B_o = \begin{pmatrix} 1\,0\,0\,0\,0\,0\,0\,0\,0\,0\,0 \\ 0\,0\,0\,1\,0\,0\,0\,0\,0\,0\,0 \end{pmatrix}^T
\tag{4.132}
$$

The gain of the disturbance observer is computed through the Kalman Filter recursion. The application of Kalman Filtering on the linearized equivalent description of the extended power system is again the so-called *Derivative-free nonlinear Kalman Filter*. Matrices, A_o, B_o and C_o are discretized using common discretization methods. This provides matrices A_d, B_d and C_d. Matrices Q and R denote the process and measurement noise covariance matrices. Matrix P is the state-vector error covariance matrix. The Kalman Filter for the state and disturbances estimation in the extended model of the power unit comprises again a *measurement update* and a *time update* stage.

measurement update:

$$
K_f(k) = P^-(k)C_d^T[C_d P^-(k)C_d^T + R]^{-1}
$$
$$
\hat{x}(k) = \hat{x}^-(k) + K_f(k)[z_m - \hat{z}_m]
\tag{4.133}
$$
$$
P(k) = P^-(k) - K_f(k)C_d P^-(k)
$$

time update:

$$P^-(k+1) = A_d P(k) A_d^T + Q$$
$$\hat{x}^-(k+1) = A_d \hat{x}(k) + B_d v(k) \tag{4.134}$$

By obtaining $\hat{z}_8 = \tilde{d}_1$ and $\hat{z}_{10} = \tilde{d}_2$, the stabilizing feedback control of the power system is modified as follows:

$$u_1 = \frac{1}{g_1(x)} [x_{1,d}^{(3)} - f_1(x) - k_1^1(\ddot{x}_1 - \ddot{x}_{1,d}) - $$
$$ -k_2^1(\dot{x}_1 - \dot{x}_1^d) - k_3^1(x_1 - x_{1,d})] - \hat{z}_8$$

$$u_2 = \frac{1}{g_2(x)} [x_{1,d}^{(4)} - f_2(x) - k_1^2(x_4^{(3)} - x_{4,d}^{(3)}) - $$
$$ -k_2^2(\ddot{x}_4 - \ddot{x}_{4,d}) - k_3^2(\dot{x}_4 - \dot{x}_4^d) - k_4^2(x_4 - x_{4,d})] - \hat{z}_{10} \tag{4.135}$$

4.2.5 STATISTICAL FAULT DIAGNOSIS WITH THE KALMAN FILTER

4.2.5.1 Fault detection

The residuals' sequence, that is the differences between (i) the real outputs of the steam turbine and synchronous generator unit and (ii) the outputs estimated by the Kalman Filter (Fig. 4.22) is a discrete error process e_k with dimension $m \times 1$ (here $m = N$ is the dimension of the output measurements vector). Actually, it is a zero-mean Gaussian white-noise process with covariance given by E_k.

A conclusion can be stated based on a measure of certainty that the power generation unit has neither been subjected to a fault nor to a cyberattack. To this end, the following *normalized error square* (NES) is defined [195]

$$\varepsilon_k = e_k^T E_k^{-1} e_k \tag{4.136}$$

The normalized error square follows a χ^2 distribution. An appropriate test for the normalized error sum is to numerically show that the following condition is met within a level of confidence (according to the properties of the χ^2 distribution)

$$E\{\varepsilon_k\} = m \tag{4.137}$$

This can be achieved using statistical hypothesis testing, which is associated with confidence intervals. A 95% confidence interval is frequently applied, which is specified using the probability region $100(1-a)$ with $a = 0.05$. Again, a two-sided probability region is considered cutting-off two end tails of 2.5% each. For M runs the normalized error square, that is obtained from the residuals sequence, is given by

$$\bar{\varepsilon}_k = \frac{1}{M} \sum_{i=1}^{M} \varepsilon_k(i) = \frac{1}{M} \sum_{i=1}^{M} e_k^T(i) E_k^{-1}(i) e_k(i) \tag{4.138}$$

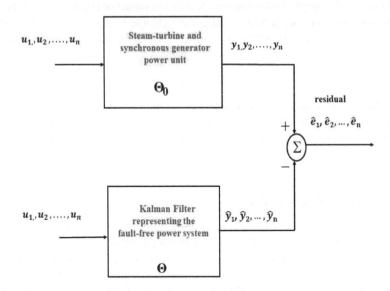

Figure 4.22 Residuals'generation for the steam-turbine and synchronous generator power unit, with the use of Kalman Filtering

where ε_i stands for the i-th run at time t_k. Then $M\bar{\varepsilon}_k$ will follow a χ^2 density with Mm degrees of freedom. This condition can be checked using a χ^2 test. The hypothesis holds, if the following condition is satisfied

$$\bar{\varepsilon}_k \in [\zeta_1, \zeta_2] \tag{4.139}$$

where ζ_1 and ζ_2 are derived from the tail probabilities of the χ^2 density.

4.2.5.2 Fault isolation

By applying the statistical test into the individual components of the thermal power system, that is the steam turbine and the synchronous generator, it is also possible to find out the specific component that has been subjected to a fault or cyberattack [195], [198]. For a power unit of n components one has to carry out n χ^2 statistical change detection tests, where each test is applied to the subset that comprises components $i-1$, i and $i+1$, $i = 1, 2, \cdots, n$. Actually, out of the n χ^2 statistical change detection tests, the ones that exhibit the highest score are those that identify the component that has been subjected to failure (the failure can be either a fault related with a change in the parameters of the electromechanical part of the power system, or can be the result of a cyberattack).

In the case of multiple faults one can identify the subset of components of the power unit that have been subjected to parametric change by applying the χ^2 statistical change detection test according to a combinatorial sequence. This means that

$$\binom{n}{k} = \frac{n!}{k!(n-k)!} \tag{4.140}$$

tests have to take place, for all clusters in the monitored power system, that finally comprise $n, n-1, n-2, \cdots, 2, 1$ components. Again the χ^2 tests that give the highest scores indicate the components which are most likely to have been subjected to failure.

4.2.6 SIMULATION TESTS

To perform the faults detection and isolation tests with the previously analyzed χ^2 statistical criterion one can define the output measurements vector $y_m = [x_1, x_2, x_4]$ where $x_1 = \theta$ (turn angle of the generator's rotor), $x_2 = \omega$ (angular speed of the synchronous generator) and $x_4 = P_m$ (mechanical power provided by the steam turbine to the synchronous generator). Thus the dimension of the outputs measurements vector is $m = 3$. Considering that the number of output vector samples is $M = 2000$ and using a 98% confidence interval for the χ^2 distribution the fault thresholds can be set as $L = 2.87$ and $U = 3.13$. In an equivalent manner when the statistical test is applied exclusively to the steam turbine the fault thresholds are defined as $L = 0.90$ and $U = 1.10$. Moreover, when the statistical test is applied only to the synchronous generator the fault thresholds are defined as $L = 1.89$ and $U = 2.11$.

In the case of detection of unknown input perturbations, the Derivative-free nonlinear Kalman Filter has been used as a state and disturbances estimator. The statistical χ^2 test has been applied as described in Chapter 5 Section 2, to the extended system that comprises both the steam turbine and the synchronous generator. By identifying in real-time the cumulative faults that affected the power system, their compensation became possible, and the feedback controller was capable of suppressing the effects of such disturbances. The obtained results are depicted in Fig. 4.23. Subsequently, the χ^2 statistical test showed that the power system remained in the normal functioning mode. However, by observing that the estimates of the additive disturbance inputs $\tilde{d}_1$ an $\tilde{d}_2$ which are provided by state variables $\hat{z}_8$ and $\hat{z}_{10}$ one can perceive the existence of perturbations that distort the power unit's control inputs. The latter results are depicted in Fig. 4.24.

In case of fault detection with χ^2 tests, the Derivative-free nonlinear Kalman Filter has been used as a state estimator and the statistical χ^2 test has been applied to the power system that comprises both the steam turbine and the synchronous generator. In particular, the following cases have been examined: (1) Fault at the control input gains of the steam turbine as shown in Fig. 4.25, (2) Fault at the control input gains of the generator as shown in Fig. 4.26, (3) fault (additive disturbance) at the control in-

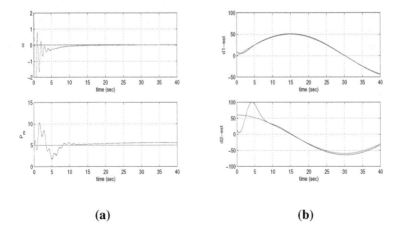

(a)   (b)

Figure 4.23 Kalman Filter-based disturbances estimator (i) convergence to reference set-points (red lines) of state variables (blue lines) of $x_2 = \omega$ (turn speed of the synchronous generator) and $x_4 = P_m$ (mechanical power provide to the generator by the steam turbine) and (ii) estimation (blue lines) by the Kalman Filter of unknown disturbance inputs (red lines)

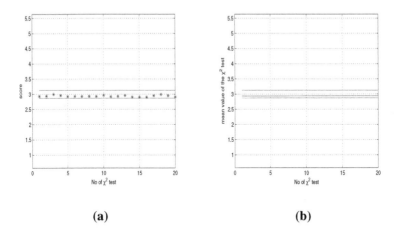

(a)   (b)

Figure 4.24 Fault diagnosis at the steam-turbine power-unit in the fault-free case (i) individual χ^2 tests and (ii) mean value of the χ^2 tests for fault diagnosis

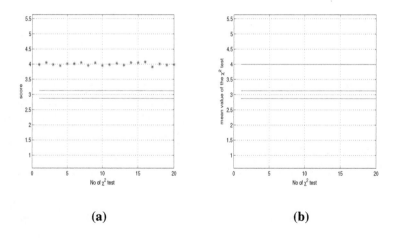

(a)   **(b)**

Figure 4.25 Fault taking place at the control input gains of the steam turbine (i) individual χ^2 tests and (ii) mean value of the χ^2 test for fault diagnosis

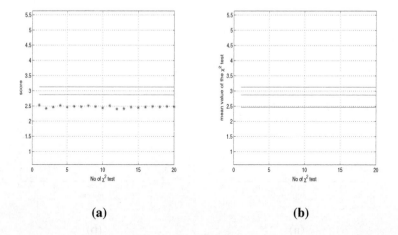

(a)   **(b)**

Figure 4.26 Fault taking place at the control input gains of the synchronous generator (i) individual χ^2 tests and (ii) mean value of the χ^2 test for fault diagnosis

put at the turbine as shown in Fig. 4.27, (4) Fault (additive disturbance) at the control input at the generator as shown in Fig 4.28, (5) Fault at the steam turbine's control input sensors as shown in Fig. 4.29, (6) Fault at the generator's control input sensors as shown in Fig. 4.30, (7) Fault at the steam turbine's output measurement sensors as shown in Fig. 4.31, (8) Fault at the generator's output measurement sensors as shown in Fig. 4.32, and (9) Fault at both the turbine's and generator's output measurement sensors as shown in Fig. 4.33.

In the case of fault isolation with χ^2 tests, again the Kalman Filter has been used as state estimator and the statistical change detection test has been carried out separately at the components of the power unit, that is the steam turbine and the power generator. The following indicative results are presented: (1) Fault isolation tests at the generator under control input disturbance at the generator, as shown in Fig. 4.34 (2) Fault isolation tests at the turbine under control input disturbance at the generator, as shown in Fig. 4.35 (3) Fault isolation tests at the generator under control input disturbance at the turbine as shown in Fig. 4.36, and (4) Fault isolation tests at the turbine under control input disturbance at the turbine, as shown in Fig. 4.37. It is confirmed that the proposed χ^2 statistical tests is efficient in isolating the faulty component of the power system.

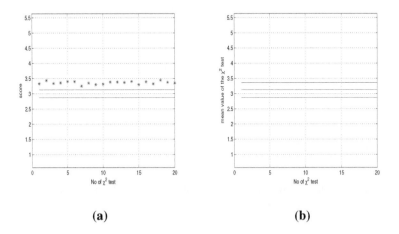

(a)   **(b)**

Figure 4.27   Fault (additive disturbance) at the control input at the turbine (i) individual χ^2 tests and (ii) mean value of the χ^2 test for fault diagnosis

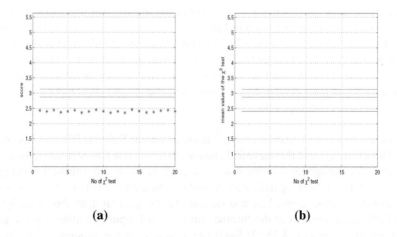

Figure 4.28 Fault (additive disturbance) at the control input at the generator (i) individual χ^2 tests and (ii) mean value of the χ^2 test for fault diagnosis

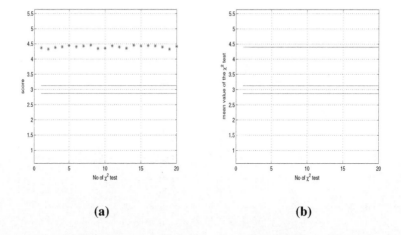

Figure 4.29 Fault at the turbine's control input sensors (i) individual χ^2 tests and (ii) mean value of the χ^2 test for fault diagnosis

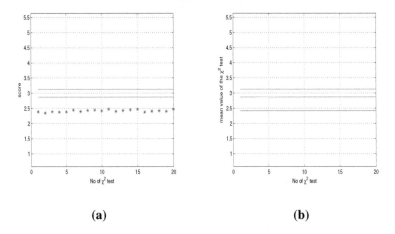

(a)   **(b)**

Figure 4.30   Fault at the generator's control input sensors (i) individual χ^2 tests and (ii) mean value of the χ^2 test for fault diagnosis

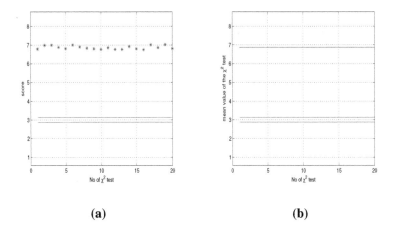

(a)   **(b)**

Figure 4.31   Fault at the turbine's output measurement sensors (i) individual χ^2 tests and (ii) mean value of the χ^2 test for fault diagnosis

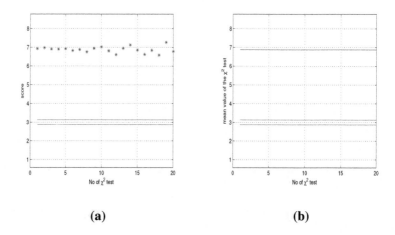

(a)   (b)

Figure 4.32 Fault at the generator's output measurement sensors (i) individual χ^2 tests and (ii) mean value of the χ^2 test for fault diagnosis

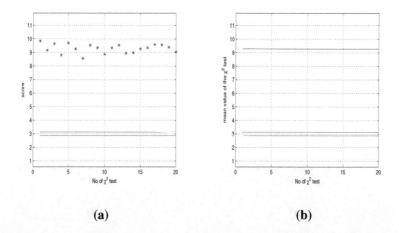

(a)   (b)

Figure 4.33 Fault at the both the generator's and the turbine's output measurement sensors (i) individual χ^2 tests and (ii) mean value of the χ^2 test for fault diagnosis

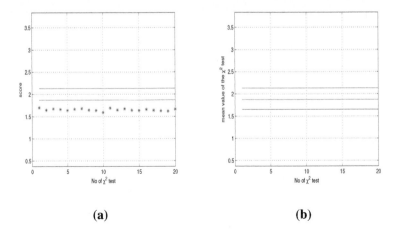

(a) **(b)**

Figure 4.34 Fault isolation at the generator for additive disturbance input affecting the generator (a) Results of the individual χ^2 tests and (b) mean value χ^2 tests

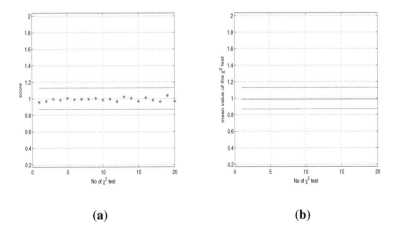

(a) **(b)**

Figure 4.35 Fault isolation at the turbine for additive disturbance input affecting the generator (a) results of the individual χ^2 tests and (b) mean value χ^2 tests

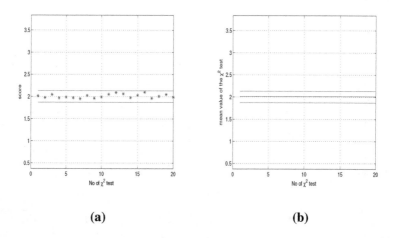

Figure 4.36   Fault isolation at the generator for additive disturbance input affecting the turbine (a) results of the individual χ^2 tests and (b) mean value χ^2 tests

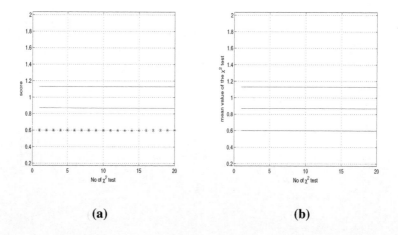

Figure 4.37   Fault isolation at the turbine for additive disturbance input affecting the turbine (a) results of the individual χ^2 tests and (b) mean value χ^2 tests

5 Fault diagnosis for wind power units and the distribution grid

5.1 FAULT DIAGNOSIS FOR WIND POWER GENERATORS

5.1.1 OUTLINE

The chapter proposes first a solution for the fault diagnosis problem of a wind power generation unit which consists of a synchronous ge], [that is actuated by a wind-turbine through a drivetrain. During the last years the need for energy has grown and the contribution of wind power generation systems to covering energy demands appears to be also continuously growing [198], [15], [219]. In particular wind power units that comprise asynchronous generators of the DFIG type are often met in power production from renewable sources. This is due to their capability to function reliably under changing operating points and at variable speeds, thus allowing also for more efficient energy harvesting [58], [127]. However, wind power generation units of the DFIG type often undergo failures of their electrical and mechanical components [5], [32]. The main reason for such failures is that wind power generation systems function under hard conditions and are exposed to various external perturbations [133], [84], [128]. Besides, by becoming part of the electricity grid and by being controlled within networked communication schemes, such types of wind power generation units are also exposed to human interventions known as cyberattacks.

The present chapter proposes a systematic method for fault diagnosis and cyberattacks detection in wind-power generation systems comprising a wind turbine, a drive train and an asynchronous generator (DFIG) [49], [211]. First, it is proven that such a power unit is differentially flat [195], [126], [230]. This means that all state variables and control inputs of the system can be written as differential functions of specific state vector elements, which are the so-called flat outputs. Moreover, the flat outputs of the system are differentially independent which means that they are not connected to each other through a relation in the form of a linear differential equation [195], [230], [245], [156]. By proving that the system is differentially flat it is also inferred that it can be transformed into an input-output linearized form and consequently it can be written into the linear canonical (Brunovsky) state-space description. For the latter representation of the wind power unit's dynamics, one can solve the related state-estimation and filtering problem [195], [75].

DOI: 10.1201/9781003527657-5

Actually, for the linearized equivalent state-space description of the power unit the filtering method, under the name Derivative-free nonlinear Kalman Filter, is applied. This filtering method consists of the standard Kalman Filter recursion applied on the linearized dynamics of the wind power unit [188], [189]. It also involves an inverse transformation related with the differential flatness properties of the system which allows for computing estimates for the state vector elements of its initial nonlinear description. The proposed Kalman Filtering method is used as a fault-free representation of the wind power unit. At a next stage, faults and attacks detection is performed by comparing the outputs of the Kalman Filter against measurements obtained from the real outputs of the wind-power unit. The differences between the outputs of the Kalman Filter and the outputs of the wind power unit form he residuals' sequence, with elements which are known to be i.i.d (independent identically distributed) and to follow the Gaussian distribution. Moreover, it is proven that the sum of the squares of the system's residuals' vectors. weighted by the inverse of the associated covariance matrix, stands for a stochastic variable which follows the χ^2 distribution [18], [189]. By exploiting the properties of the χ^2 distribution, a statistical test can be formulated for performing condition monitoring of the wind-power system, with certainty levels of the order of 96% or 98% [198], [18], [195].

5.1.2 DYNAMIC MODEL OF THE WIND POWER UNIT

Dynamics of the mechanical part of the wind power unit:

The dynamic model of the wind power generation system is shown in Fig. 5.1. The rotational motion of the wind-turbine and of the drivetrain system is described by the following differential equation:

$$J_t \frac{d\omega_t}{dt} = c_{ba} T_m - T_{ls} - B_t \omega_t \qquad (5.1)$$

where ω_t is the turn speed of the wind turbine, T_m is the mechanical torque provided by the wind, c_{ba} is a coefficient related with the pitch angle of the turbine's blades and regulates the effect of the mechanical torque on the wind turbine. Moreover, T_{ls} is the torque of the low-speed shaft of the drivetrain (shaft on the side of the wind turbine), B_t is a damping coefficient (friction) that opposes to the rotational motion of the shaft. The torque of the low-speed shaft is

$$T_{ls} = K_1(\theta_t - \theta_g) + D_1(\omega_t - \omega_g) \qquad (5.2)$$

where θ_t is the turn angle of the wind turbine, θ_g it the turn angle of the generator's (DFIG) rotor, K_1 is an elasticity coefficient and D_1 is a damping coefficient. In Eq. (5.2) the effects of the damping term $D_1(\omega_t - \omega_g)$ in the torque of the low speed shaft can be considered as negligible, thus one can use finally $T_{ls} = K_1(\theta_t - \theta_g)$. Taking into account the preservation of mechanical power, the relation between the torque of the wind turbine's shaft (low-speed shaft) and the torque of the DFIG shaft

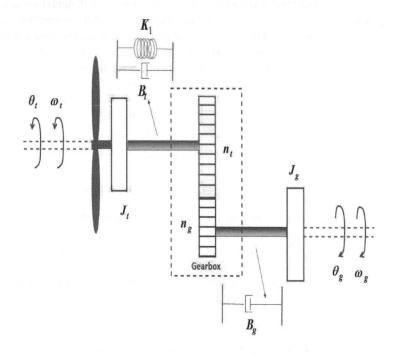

Figure 5.1   Wind power generation unit comprising a multi-mass drive-train and a DFIG

(high-speed shaft) is given as follows:

$$\frac{T_t}{T_g} = \frac{T_{ls}}{T_{hs}} = \frac{\omega_g}{\omega_t} = \frac{n_t}{n_g} \tag{5.3}$$

where n_t is the number of teeth at the gear placed at the wind turbine's side, while n_g is the number of teeth at the gear placed at the generator's side. The rotational motion of the Doubly-Fed Induction Generator is described by the following differential equation:

$$J_g \frac{d\omega_g}{dt} = T_{hs} - T_e - B_g \omega_g \tag{5.4}$$

where J_g is the moment of inertia of the generator's rotor, ω_g is the turn speed of the generator's rotor, T_{hs} is the torque exerted at the high-speed shaft of the generator, and B_g is a damping coefficient (friction) that opposes to the rotor's motion. Moreover, T_e is the electromagnetic torque of the generator that inhibits the rotor's motion and is given by

$$T_e = \eta \left(i_{r_q} \psi_{s_d} - i_{r_d} \psi_{s_q} \right) \tag{5.5}$$

where η is a variable related with the mutual inductance coefficient (between the generator's stator and rotor) and with the number of poles of the generator, $[i_{rd}, i_{rq}]$ are the vector components of the DFIG's rotor currents expressed in the asynchronously rotating dq reference frame and $[\psi_{s_d}, \psi_{s_q}]$ are the vector components of the DFIG's stator magnetic flux also expressed in the dq reference frame. About the high-speed shaft torque at the side of the generator, it holds that

$$
\begin{aligned}
T_{hs} = T_{ls}\frac{\omega_t}{\omega_g} \Rightarrow T_{hs} = T_{ls}\frac{n_g}{n_t} \\
T_{hs} = K_1(\theta_t - \theta_g)\frac{n_g}{n_t}
\end{aligned}
\tag{5.6}
$$

By substituting Eq. (5.1) into Eq. (5.4) one gets

$$
J_t\frac{d\omega_t}{dt} = c_{ba}T_m - K_1(\theta_t - \theta_g) - B_t\omega_t
\tag{5.7}
$$

By substituting Eq. (5.6) and Eq. (5.5) into Eq. (5.3) one gets

$$
J_g\frac{d\omega_g}{dt} = K_1(\theta_t - \theta_g)\frac{n_t}{n_g} - \eta(i_{r_q}\psi_{s_d} - i_{r_d}\psi_{s_q}) - B_g\omega_g
\tag{5.8}
$$

As a result of the above, the joint dynamics of the drivetrain and of the DFIG is expressed by the following two differential equations [198]:

$$
\begin{aligned}
J_t\frac{d\omega_t}{dt} = c_{ba}T_m - K_1(\theta_t - \theta_g) - B_t\omega_t \\
J_g\frac{d\omega_g}{dt} = K_1(\theta_t - \theta_g)\frac{n_t}{n_g} - \eta(i_{r_q}\psi_{s_d} - i_{r_d}\psi_{s_q}) - B_g\omega_g
\end{aligned}
\tag{5.9}
$$

Next, by defining the state variables $x_1 = \theta_t$, $x_2 = \omega_t$, $x_3 = \theta_g$, and $x_4 = \omega_g$ and after using Eq. (5.9) one arrives at the following state-space description of the mechanical part of the wind power generation unit:

$$
\begin{aligned}
\dot{x}_1 &= x_2 \\
\dot{x}_2 &= \frac{c_{ba}}{J_t}T_m - \frac{K_1}{J_t}(\theta_t - \theta_g) - \frac{B_t}{J_t}\omega_t \\
\dot{x}_3 &= x_4 \\
\dot{x}_4 &= \frac{K_1}{J_g}(x_1 - x_3)\frac{n_t}{n_g} - \frac{\eta}{J_g}(x_8x_5 - x_7x_6) - \frac{B_g}{J_g}x_4
\end{aligned}
\tag{5.10}
$$

Dynamics of the electrical part of the wind power unit:

In a compact form and under the field-orientation assumption, the doubly-fed induction generator can be described by the following set of equations in the dq reference frame that rotates at a speed denoted as ω_{dq} [198]:

$$
\begin{aligned}
\frac{d\psi_{sd}}{dt} &= \omega_{dq}\psi_{sq} - \frac{1}{\tau_s}\psi_{sd} + \frac{M}{\tau_s}i_{rd} + v_{sd} \\
\frac{d\psi_{sq}}{dt} &= -\frac{1}{\tau_s}\psi_{sq} - \omega_{dq}\psi_{sd} + \frac{M}{\tau_s}i_{rq} + v_{sq} \\
\frac{di_{rd}}{dt} &= -\beta\omega_g\psi_{sq} + \frac{\beta}{\tau_s}\psi_{sd} + (\omega_{dq} - \omega_g)i_{rq} - \gamma_2 i_{rd} - \beta v_{sd} + \frac{1}{\sigma L_r}v_{rd} \\
\frac{di_{rq}}{dt} &= \frac{\beta}{\tau_s}\psi_{sq} + \beta\omega_g\psi_{sd} - \gamma_2 i_{rq} - (\omega_{dq} - \omega_g)i_{rd} - \beta v_{sq} + \frac{1}{\sigma L_r}v_{rq}
\end{aligned}
\tag{5.11}
$$

where ψ_{s_q}, ψ_{s_d}, i_{r_q}, i_{r_d} are the stator flux and the rotor currents, v_{s_q}, v_{s_d}, v_{r_q}, v_{r_d} are the stator and rotor voltages, L_s and L_r are the stator and rotor inductances, ω_r is the rotor's angular velocity, M is the mutual inductance (between the stator and the rotor). Moreover, denoting as R_s and R_r the stator and rotor resistances the following parameters are defined $\sigma = 1 - \frac{M^2}{L_r L_s}$, $\beta = \frac{1-\sigma}{M\sigma}$, $\tau_s = \frac{L_s}{R_s}$, $\tau_r = \frac{L_r}{R_r}$. The dynamic model of the electric part of the doubly-fed induction generator can be also written in state space equations form by defining the following state variables: $x_5 = \psi_{s_d}$, $x_6 = \psi_{s_q}$, $x_7 = i_{r_d}$ and $x_8 = i_{r_q}$. It holds that

$$
\begin{aligned}
\dot{x}_5 &= -\tfrac{1}{\tau_s}x_5 + \omega_{dq}x_6 + \tfrac{M}{\tau_s}x_7 + v_{sd} \\
\dot{x}_6 &= -\omega_{dq}x_5 - \tfrac{1}{\tau_s}x_6 + \tfrac{M}{\tau_s}x_8 + v_{sq} \\
\dot{x}_7 &= -\beta x_4 x_6 + \tfrac{\beta}{\tau_s}x_5 + (\omega_{dq} - x_4)x_8 - \gamma_2 x_7 - \beta v_{s_d} + \tfrac{1}{\sigma L_r}v_{rd} \\
\dot{x}_8 &= \tfrac{\beta}{\tau_s}x_6 + \beta x_4 x_5 - \gamma_2 x_8 - (\omega_{dq} - x_4)x_7 - \beta v_{s_q} + \tfrac{1}{\sigma L_r}v_{rq}
\end{aligned}
\tag{5.12}
$$

Joint electromechanical dynamics of the wind power system: Using Eq. (5.10) and Eq. (5.12) and after defining the control inputs $u_1 = c_{ba}$ (term related with the pitch angle of the turbine's blades), $u_2 = v_{r_d}$ (d-axis component of control input voltage applied to the DFIG's rotor) and $u_3 = v_{r_q}$ (q-axis component of control input voltage applied to the DFIG's rotor), one arrives at the following state-space equations which describe jointly the dynamics of the mechanical and electrical part of the power generation unit:

$$
\dot{x} = f(x) + g(x)u \tag{5.13}
$$

where $x \in R^{8 \times 1}$, $f(x) \in R^{8 \times 1}$, $g(x) \in R^{8 \times 3}$ and $u \in R^{3 \times 1}$. In particular, matrices f(x) and g(x) are defined as follows:

$$
f = \begin{pmatrix}
x_2 \\
-\frac{K_1}{J_t}(\theta_t - \theta_g) - \frac{B_t}{J_t}\omega_t \\
x_4 \\
\frac{K_1}{J_g}(x_1 - x_3)\frac{n_t}{n_g} - \frac{\eta}{J_g}(x_8 x_5 - x_7 x_6) - \frac{B_g}{J_g}x_4 \\
-\frac{1}{\tau_s}x_5 - \omega_{dq}x_6 + \frac{M}{\tau_s}x_8 + v_{sq} \\
-\beta x_4 x_6 + \frac{\beta}{\tau_s}x_5 + (\omega_{dq} - x_4)x_8 - \gamma_2 x_7 - \beta v_{s_d} \\
\frac{\beta}{\tau_s}x_6 + \beta x_4 x_5 - \gamma_2 x_8 - (\omega_{dq} - x_4)x_7 - \beta v_{s_q}
\end{pmatrix}
\tag{5.14}
$$

$$
g = \begin{pmatrix}
0 & 0 & 0 \\
\frac{T_m}{J_t} & 0 & 0 \\
0 & 0 & 0 \\
0 & 0 & 0 \\
0 & 0 & 0 \\
0 & 0 & 0 \\
0 & \frac{1}{\sigma L_r} & 0 \\
0 & 0 & \frac{1}{\sigma L_r}
\end{pmatrix}
\tag{5.15}
$$

5.1.3   PROOF OF THE DIFFERENTIAL FLATNESS PROPERTIES OF THE WIND POWER GENERATION SYSTEM

Next, it will be proven that the wind-power system which comprises the wind-turbine, the drivetrain and the DFIG is a differentially flat one. This will allow for solving the state estimation problem of the power unit. The state-space model of Eq. (5.13) is used. The flat outputs vector is taken to be $Y = [x_1, x_3, x_5]^T = [\theta_t, \theta_g, \psi_{s_d}]^T$ where θ_t is the turn angle of the wind turbine, θ_g is the turn angle of the DFIG's rotor and ψ_{s_d} is the component of the magnetic flux axis along the d-axis of the dq reference frame. From the first row of the state-space model one has:

$$x_2 = \dot{x}_1 \Rightarrow x_2 = h_2(Y, \dot{Y}) \tag{5.16}$$

that is x_2 is a differential function of the flat outputs of the system. From the third row of the state-space model, one obtains

$$x_4 = \dot{x}_3 \Rightarrow x_4 = h_4(Y, \dot{Y}) \tag{5.17}$$

which signifies that x_4 is also a differential function of the flat outputs of the system. From the sixth row of the state-space model one has that due to the assumption of field orientation, that is $x_6 = \psi_{s_q} = 0$, the aforementioned row becomes an algebraic equation. Thus, this row will not be taken into account in the proof of the differential flatness properties of the power unit. From the fifth row of the state-space model and using that $x_6 = \psi_{s_q} = 0$ and that $v_{s_d} = V_s$ one can solve with respect to x_7. This gives

$$x_7 = \frac{\tau_s}{M}\dot{x}_5 + \omega_{d_q}x_5 - \frac{1}{\tau_s}x_6 - v_{s_q} \Rightarrow x_7 = h_7(Y, \dot{Y}) \tag{5.18}$$

which means that state variable x_7 is a differential function of the flat output of the system. From the fourth row of the state-space model one can solve with respect to x_8. By denoting $L_m = \frac{\eta}{J_g}$, this gives

$$\dot{x}_4 = -\frac{K_1}{J_g}(x_1 - x_3)\frac{n_g}{n_t} + L_m x_7 x_6 + \frac{B_g}{J_g}x_4 - L_m x_5 x_8 \Rightarrow$$

$$x_8 = \frac{1}{L_m x_5}[\dot{x}_4 - \frac{K_1}{J_g}(x_1 - x_3)\frac{n_g}{n_t} + L_m x_7 x_6 + \frac{B_g}{J_g}x_4] \Rightarrow \tag{5.19}$$

$$x_8 = h_8(Y, \dot{Y})$$

Consequently x_8 is also a function of the differentially flat outputs of the system. Next, from the second row of the state-space model of the system one can solve with respect to the control input u_1. This gives for $u_1 = c_{ba}$:

$$u_1 = \frac{J_1}{T_m}[\dot{x}_2 + \frac{K_1}{J_t}(x_1 - x_3) + \frac{B_t}{J_t}x_2] \Rightarrow u_1 = q_1(Y, \dot{Y}) \tag{5.20}$$

which means that the control input u_1 is a differential function of the flat outputs of the system. Equivalently from the seventh row of the state-space model of the system one can solve with respect to the control input u_2. This gives:

$$u_2 = \sigma L_r \{\dot{x}_7 + \beta x_4 x_6 + \frac{\beta}{\tau_s}x_5 + (\omega_{dq} - x_4)x_8 +$$

$$+ \gamma_2 x_7 + \beta v_{s_d}\} \Rightarrow u_2 = q_2(Y, \dot{Y}) \tag{5.21}$$

which shows that the control input u_2 is a differential function of the flat outputs of the system. Finally, from the eight row of the state-space model of the system one can solve with respect to the control input u_3. This gives:

$$u_3 = \sigma L_r \{ \dot{x}_8 + \tfrac{\beta}{\tau_s} x_6 - \beta x_4 x_5 - (\omega_{dq} - x_4)x_7 + \\ + \gamma_2 x_8 + \beta v_{s_q} \} \Rightarrow u_3 = q_3(Y, \dot{Y}) \tag{5.22}$$

Considering that due to field orientation $x_6 = 0$ and $v_{s_q} = 0$ one has that the control u_3 is also written as a differential function of the flat outputs of the power unit. As a result of the previous analysis, all state variables and the control inputs of the wind power generation unit can be written as differential functions of the flat outputs vector $Y = [x_1, x_3, x_5]^T = [\theta_t, \theta_g, \psi_{s_d}]^T$. Consequently, the wind power generation system is a differentially flat one.

5.1.4 TRANSFORMATION OF THE WIND-POWER SYSTEM INTO AN INPUT-OUTPUT LINEARIZED FORM

The flat outputs of the system are differentiated successively, until the control inputs reappear. Thus, from the first row of the state-space model of Eq. (5.13) to Eq. (5.15) and using that $\dot{x}_1 = x_2$, one obtains:

$$\ddot{x}_1 = -\frac{K_1}{J_1}(x_1 - x_3) - \frac{B_t}{J_t}x_2 + \frac{T_m}{J_t}u_1 \tag{5.23}$$

Equivalently, one can write

$$\ddot{x}_1 = f_a(x) + g_{a_1}(x)u_1 + g_{a_2}(x)u_2 + g_{a_3}(x)u_3 \tag{5.24}$$

$$f_a = -\frac{K_1}{J_1}(x_1 - x_3) - \frac{B_t}{J_t}x_2 \tag{5.25}$$

$$g_{a_1} = \frac{T_m}{J_t} \quad g_{a_2} = 0 \quad g_{a_3} = 0 \tag{5.26}$$

Next, from the third row of the state-space model of Eq. (5.13) to Eq. (5.15), one obtains

$$\ddot{x}_3 = x_4 \Rightarrow \dddot{x}_3 = \dot{x}_4 \Rightarrow \\ \dddot{x}_3 = \frac{K_1}{J_g}(x_1 - x_3)\frac{n_g}{n_t} - \frac{\eta}{J_g}(x_8 x_5 - x_7 x_6) - \frac{B_g}{J_g}x_4 \tag{5.27}$$

Due to the field-orientation condition, it holds that $x_6 = \psi_{s_q} = 0$. Consequently, one has

$$x_3^{(3)} = f_b(x) + g_{b_1}(x)u_1 + g_{b_2}(x)u_2 + g_{b_3}(x)u_3 \tag{5.28}$$

$$f_b = \frac{K_1}{J_g}(x_2 - x_4)\frac{n_g}{n_t} - \frac{\eta}{J_g}x_5[\tfrac{\beta}{\tau_s}x_6 + \\ + \beta x_4 x_5 + (\omega_{dq} - x_4)x_7 - \gamma_2 x_8 \beta v_{s_q}] - \\ - \frac{\eta}{J_g}x_8[-\omega_{dq}x_5 - \tfrac{1}{\tau_s}x_6 + \tfrac{M}{\tau_s}x_8 + v_{s_q}] - \\ - \frac{B_g}{J_g}[\frac{K_1}{J_g}(x_1 - x_3)\frac{n_g}{n_t} - \frac{\eta}{J_g}(x_8 x_5 - x_7 x_6) - \frac{B_g}{J_g}x_4] \tag{5.29}$$

$$g_{b_1} = 0 \quad g_{b_2} = 0 \quad g_{b_3} = -\frac{\eta}{J_g} x_5 \left(\frac{1}{\sigma L_s} \right) \tag{5.30}$$

Next, from the fifth row of the state-space model of Eq. (5.13) to Eq. (5.15), one obtains:

$$\ddot{x}_5 = f_c(x) + g_{c_1}(x) u_1' + g_{c_2}(x) u_2 + g_{c_3}(x) u_3 \tag{5.31}$$

$$f_c = \left\{ \left[\frac{1}{\tau_s^2} x_5 - \frac{M}{\tau_s^2} x_7 \right] + \frac{M}{\tau_s} \left[\frac{\beta}{\tau_s} x_5 + (\omega_{dq} - x_4) x_8 - \gamma_2 x_7 - \beta v_{s_d} \right] \right\} \tag{5.32}$$

$$g_{c_1} = 0 \quad g_{c_2} = \frac{M}{\tau_s} \left(\frac{1}{\sigma L_s} \right) \quad g_{c_3} = 0 \tag{5.33}$$

Using the previous analysis, the input-output linearized description of the system becomes:

$$\begin{pmatrix} \ddot{x}_1 \\ x_3^{(3)} \\ \ddot{x}_5 \end{pmatrix} = \begin{pmatrix} f_a(x) \\ f_b(x) \\ f_c(x) \end{pmatrix} + \begin{pmatrix} g_{a_1}(x) & g_{a_2}(x) & g_{a_3}(x) \\ g_{b_1}(x) & g_{b_2}(x) & g_{b_3}(x) \\ g_{c_1}(x) & g_{c_2}(x) & g_{c_3}(x) \end{pmatrix} \begin{pmatrix} u_1 \\ u_2 \\ u_3 \end{pmatrix} \tag{5.34}$$

or equivalently by defining the vectors $X = [\dot{x}_1, \ddot{x}_3, \dot{x}_5]^T$, and $U = [u_1, u_2, u_3]^T$ one has

$$\dot{X} = F(x) + G(x) U \tag{5.35}$$

Next one can define the following virtual control inputs $v_1 = f_a(x) + g_{a_1}(x) u_1 + g_{a_2}(x) u_2 + g_{a_3}(x) u_3$, $v_2 = f_b(x) + g_{b_1}(x) u_1 + g_{b_2}(x) u_2 + g_{b_3}(x) u_3$ and $v_3 = f_c(x) + g_{c_1}(x) u_1 + g_{c_2}(x) u_2 + g_{c_3}(x) u_3$. The input-output linearized form of the system is written as:

$$\ddot{x}_1 = v_1 \quad x_3^{(3)} = v_2 \quad \ddot{x}_5 = v_1 \tag{5.36}$$

Using Eq. (36), and by defining the reference trajectories (setpoints) $x_{1,d}$, $x_{3,d}$ and $x_{5,d}$, the stabilizing feedback control of the wind power generation unit is

$$v_1 = \ddot{x}_{1,d} - k_1^a(\dot{x}_1 - \dot{x}_1^d) - k_2^a(x_1 - x_{1,d})$$

$$v_2 = x_{3,d}^{(3)} - k_1^b(\ddot{x}_3 - \ddot{x}_{3,d}) - k_2^b(\dot{x}_3 - \dot{x}_{3,d}) - k_3^b(x_3 - x_{3,d}) \tag{5.37}$$

$$v_3 = \ddot{x}_{5,d} - k_1^c(\dot{x}_5 - \dot{x}_{5,d}) - k_2^c(x_5 - x_{5,d})$$

5.1.5 STATE ESTIMATION WITH KALMAN FILTERING

5.1.5.1 State estimation

Using the input-output linearized model of Eq. (36) the state estimator for the wind-power generation system is given by

$$\dot{\hat{Z}} = A_m \hat{Z} + B_m V + K(Z_m - C_m \hat{Z})$$

$$\hat{Z}_m = C_m \hat{Z} \tag{5.38}$$

where $Z = [\theta_t, \dot{\theta}_t, \theta_g, \dot{\theta}_g, \psi_{s_d}, \dot{\psi}_{s_d}, \ddot{\psi}_{s_d}]^T$, the estimator's gain is $K \in R^{7 \times 2}$, and

$$
A = \begin{pmatrix} 0\,1\,0\,0\,0\,0\,0 \\ 0\,0\,0\,0\,0\,0\,0 \\ 0\,0\,0\,1\,0\,0\,0 \\ 0\,0\,0\,0\,1\,0\,0 \\ 0\,0\,0\,0\,0\,0\,0 \\ 0\,0\,0\,0\,0\,0\,1 \\ 0\,0\,0\,0\,0\,0\,0 \end{pmatrix} \quad B = \begin{pmatrix} 0\,0\,0 \\ 1\,0\,0 \\ 0\,0\,0 \\ 0\,0\,0 \\ 0\,1\,0 \\ 0\,0\,0 \\ 0\,0\,1 \end{pmatrix} \quad C = \begin{pmatrix} 1\,0\,0\,0\,0\,0\,0 \\ 0\,0\,1\,0\,0\,0\,0 \\ 0\,0\,0\,0\,1\,0\,0 \end{pmatrix} \tag{5.39}
$$

The proposed control scheme can work with the use of measurements from a small number of sensors. That is, there is need to obtain measurements of only $y_1 = \theta_t$ which is the turn angle of the turbine, of $y_2 = \theta_g$ which is the turn angle of the generator's rotor, and of the magnetic flux $y_3 = \psi_s = \sqrt{\psi_{s_d}^2 + \psi_{s_q}^2}$, which due to the orientation of the magnetic field becomes $y_3 = \psi_s = \psi_{s_d}$. Since the stator flux ψ_s cannot be measured directly from a sensor (e.g. the use of Hall sensors in an electric machine with a rotating part would not be efficient), one can use the relation between the stator flux and the stator and rotor currents to calculate ψ_s. Thus one has:

$$
\begin{aligned} \psi_{s_d} &= L_s i_{s_d} + M i_{r_d} \\ \psi_{s_q} &= 0 \end{aligned} \tag{5.40}
$$

which means that by measuring stator and rotor currents one can obtain indirectly the stator's magnetic flux ψ_{s_d} [1]. Next, one can compute the dynamics of the magnetic flux, jointly with the dynamics of the rotor, through the use of the Derivative-free Nonlinear Kalman Filter. The observer's gain K_f is obtained through the Kalman Filter recursion. The use of the Kalman Filter in the linearized system is known as Derivative-free nonlinear Kalman Filter. The estimation process is complemented by inverse transformations relying on the differential flatness properties of the system. These allow for identifying the state variables of the initial nonlinear system. Matrices A_o, B_o and C_o are substituted by their discrete-time equivalents, through the application of common discretization procedures. These equivalents are denoted as: A_d, B_d and C_d. Thus, the Kalman Filter's is:

measurement update:

$$
\begin{aligned} K_f(k) &= P^-(k) C_d^T [C_d P^-(k) C_d^T + R(k)]^{-1} \\ \hat{X}(k) &= \hat{X}^-(k) + K_f(k)[Z_m(k) - \hat{Z}_m(k)] \\ P(k) &= P^-(k) - K(k) C_d P^-(k) \end{aligned} \tag{5.41}
$$

time update:

$$
\begin{aligned} P^-(k+1) &= A_d P(k) A_d^T + Q(k) \\ \hat{X}^-(k+1) &= A_d \hat{x}(k) + B_d U(k) \end{aligned} \tag{5.42}
$$

where $P^-(k)$ is the state vector's error covariance matrix before receiving $Z_m(k)$, $P(k)$ is the state vector's error covariance matrix after receiving $Z_m(k)$ at sampling instant k, $Q(k)$ is the process noise covariance matrix and $R(k)$ is the measurement noise covariance matrix.

5.1.5.2 Fault detection and isolation

The residuals' sequence, that is the differences between (i) the real outputs of the wind turbine and asynchronous generator power unit and (ii) the outputs estimated by the Kalman Filter (Fig. 5.2) is a discrete error process e_k with dimension $m \times 1$ (here $m = N$ is the dimension of the output measurements vector). Actually, it is a zero-mean Gaussian white-noise process with covariance given by E_k. A conclusion can be stated based on a measure of certainty that the power generation unit has neither been subjected to a fault nor to a cyberattack.

To this end, the following normalized error square (NES) is defined [195]

$$\varepsilon_k = e_k^T E_K^{-1} e_k \qquad (5.43)$$

The normalized error square follows a χ^2 distribution. An appropriate test for the normalized error sum is to numerically show that the condition $E\{e_k\} = m$ is met within a level of confidence (according to the properties of the χ^2 distribution). This can be achieved using statistical hypothesis testing, which is associated with confi-

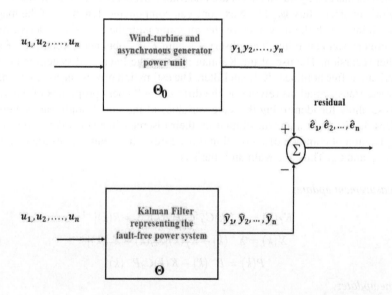

Figure 5.2 Residuals'generation for the wind-turbine and asynchronous generator power unit, with the use of Kalman Filtering

dence intervals. A 95% confidence interval is frequently applied, which is specified using the probability region $100(1 - a)$ with $a = 0.05$. Actually, a two-sided probability region is considered cutting-off two end tails of 2.5% each. For M runs the normalized error square, that is obtained from the residuals' sequence, is given by

$$\bar{\varepsilon}_k = \frac{1}{M}\sum_{i=1}^{M}\varepsilon_k(i) = \frac{1}{M}\sum_{i=1}^{M}e_k^T(i)E_k^{-1}(i)e_k(i) \tag{5.44}$$

where ε_i stands for the i-th run at time t_k. Then $M\bar{\varepsilon}_k$ will follow a χ^2 density with Mm degrees of freedom. This condition can be checked using a χ^2 test. The hypothesis holds, if the condition $\bar{\varepsilon}_k \in [\zeta_1, \zeta_2]$ is satisfied, where ζ_1 and ζ_2 are derived from the tail probabilities of the χ^2 density. By applying the statistical test into the individual components of the wind-power system, that is the wind-turbine and the asynchronous generator (DFIG), it is also possible to find out the specific component that has been subjected to a fault or cyberattack [1], [17].

5.1.6   SIMULATION TESTS

To perform the faults detection and isolation tests with the previously analyzed χ^2 statistical criterion one can define the output measurements vector $y_m = [x_1, x_3, x_5]$. Thus the dimension of the outputs measurements vector is $m = 3$. Considering that the number of output vector samples is $M = 2000$ and using a 98% confidence interval for the χ^2 distribution the fault thresholds can be as $L = 2.87$ and $U = 3.13$. In an equivalent manner when the statistical test is applied exclusively to the wind turbine the fault thresholds are defined as $L = 0.90$ and $U = 1.10$. Moreover, when the statistical test is applied only to the DFIG asynchronous generator the fault thresholds are defined as $L = 1.89$ and $U = 2.11$. Indicative results are shown in Fig. 5.3 to Fig. 5.5

5.2   FAULT DIAGNOSIS FOR THE ELECTRIC POWER DISTRIBUTION GRID

5.2.1   INTRODUCTION

At a second stage, the chapter proposes a solution for the fault diagnosis problem of the PMU sensors that monitor currents and voltages in the power distribution grid. The electricity grid is a spatially distributed and complicated infrastructure, which is not only subjected to faults due to harsh operating conditions and aging but is also exposed to attacks from external intruders who might attempt to cause malfunctioning or permanent damage to its constituents [29], [176], [22], [89], [62]. Thus, there is need for systematic methods that will detect and isolate attacks in specific components of the power grid such as AC or DC power generation units, transformers, converters, inverters and active power filters, power storage elements, or other equipment related to power generation, transmission and distribution [102], [121], [236],

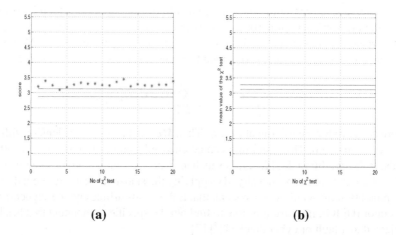

Figure 5.3 Test case 1: (a) values of successive χ^2 tests performed for the entire wind-power unit under fault affecting the inputs gain matrix G of Eq. (5.35) at the generator's part (b) mean value of the individual χ^2 tests under fault affecting the input gains matrix G at the generator's part

[150], [42]. One can also consider methods for detecting and deterring malicious attacks that target the computing and data acquisition units, the communication links between various devices in the power grid or measurement units and sensors.

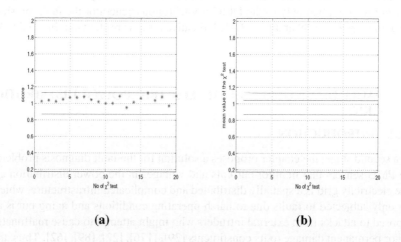

Figure 5.4 Test case 1: (a) values of successive χ^2 tests performed for the wind-turbine under fault affecting the input gains matrix G of Eq. (5.35) at the generator's part (b) mean value of the individual χ^2 tests under fault affecting the input gains matrix G at the generator's part

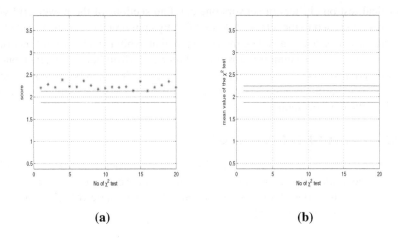

(a)   (b)

Figure 5.5 Test case 1: (a) values of successive χ^2 tests performed for the DFIG under fault affecting the input gains matrix G of Eq. (5.35) at the generator's part (b) mean value of the individual χ^2 tests under fault affecting the input gains matrix G at the generator's part

In the present chapter a computational method based on Kalman Filtering and statistical decision making is developed for detecting and isolating attacks to sensors of the power grid, such as phasors. On the one side, the Kalman Filter has been widely used in modeling and estimation tasks for the power grid [145], [182], [61], [146], [101]. On the other side statistical decision making becomes a necessity in the condition monitoring of the power grid so as to avoid false alarms and to capture incipient faults [180], [229], [122], [161], [235]. The Kalman Filter, makes use of a fault-free reference model of the sensors. It actually stands for a virtual sensor that emulates the real sensors functioning in normal conditions [191], [192], [195], [199]. In this section, by comparing the output of the Kalman Filter against the output of the real sensors a differences' sequence known as residuals is generated. The processing of the residuals with the use of statistical decision making criteria provides an indication about the existence of parametric changes in the sensors' model [18], [189], [104], [239], [270], [20]. The square of the residuals vector weighted by the inverse of the associated covariance matrix is known to follow the χ^2 statistical distribution. The weighted residuals' vector square provides also a statistical test for determining if the sensors of the power grid have undergone an attack [52], [79].

By applying a confidence intervals approach based on the χ^2 distribution one can find thresholds against which the values of the previous statistical test is compared. As long as the value of the statistical test remains between these thresholds one can conclude (up to a certain confidence value e.g. 98%) that the functioning of the grid's sensors remains normal. Otherwise one can infer that the grid's sensors have been exposed to a failure or an attack by an external intruder. Moreover, by applying the statistical test on clusters of sensors one can find sections of the power grid which have been exposed to the attack. Additionally, by applying the statistical test at each individual sensor one can isolate those sensors that exhibit an abnormal functioning. Furthermore, by redesigning the Kalman Filter as a disturbance observer one can identify additive disturbances to the sensors' output, which can be an indication that the sensor's malfunctioning has been the result of the intruders attack and not the result of the device's fault.

5.2.2   MODEL OF THE POWER GRID

The model of the three-area three-machine power grid is depicted in Fig. 5.6. Voltage measurements are obtained with the use of phasors at each bus connecting the power generation units. . At each measurement point, the three-phase voltage signals of the grid are recorded, that is [145]

$$V_1^i = A_1^i cos(\omega t + \phi)$$
$$V_2^i = A_2^i cos(\omega t + \phi - \tfrac{2\pi}{3})$$
$$V_3^i = A_3^i cos(\omega t + \phi - \tfrac{4\pi}{3})$$
(5.45)

Without loss of generality, the voltage V_1^i of the first phase is retained for further processing from measurement point i. This is written as

$$V_1^i = A_1^i cos(\omega t + \phi) \Rightarrow$$
$$V_1^i = A_1^i [cos(\omega t)cos(\phi) - sin(\omega t)sin(\phi)]$$
(5.46)

Next, by defining the state variables

$$x_1^i = A_1^i cos(\phi) \quad x_2^i = A_1^i sin(\phi)$$
(5.47)

and using Eq. (5.46) one has that the voltage recorded at the i-th measurement node is given by

$$V_1^i = x_1^i cos(\omega t) - x_2^i sin(\omega t)$$
(5.48)

In the case of the three-area three-machine electric power system that is depicted in Fig. 5.6, the voltages monitored at the three measurement units are given by

$$V_1^1 = x_1^1 cos(\omega t) - x_2^1 sin(\omega t)$$
$$V_1^2 = x_1^2 cos(\omega t) - x_2^2 sin(\omega t)$$
$$V_1^3 = x_1^3 cos(\omega t) - x_2^3 sin(\omega t)$$
(5.49)

The power system's state vector is taken to be $x = [x_1^1, x_2^1, x_1^2, x_2^2, x_1^3, x_2^3]$. The state-space description of the system is in the form

$$\dot{x} = Ax + w$$
$$y = Cx + v$$
(5.50)

where $x \in R^{6 \times 1}$ is the previously defined state vector, $A = 0 \in R^{6 \times 6}$ is a matrix having all its elements equal to zero, $w \in R^{6 \times 1}$ is a process noise which is a white Gaussian

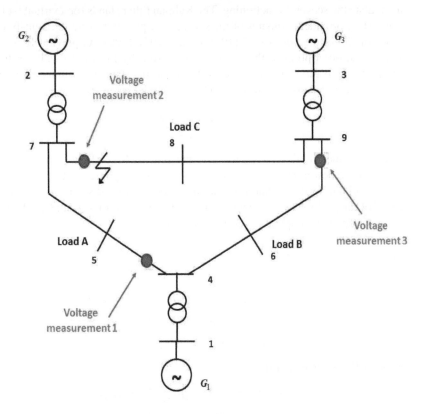

Figure 5.6   The model of the three-area three-machine power grid. Voltage measurements are obtained with the use of phasors at each bus connecting the power generation units

signal, $y \in R^{3 \times 1}$ is the measured outputs vector, $C \in R^{3 \times 3}$ is the measurement matrix and $v \in R^{3 \times 1}$ is the measurement noise which is also a white Gaussian signal.

In particular, about the measurement matrix C it holds

$$C = \begin{pmatrix} cos(\omega t) & -sin(\omega t) & 0 & 0 & 0 & 0 \\ 0 & 0 & cos(\omega t) & -sin(\omega t) & 0 & 0 \\ 0 & 0 & 0 & 0 & cos(\omega t) & -sin(\omega t) \end{pmatrix} \quad (5.51)$$

5.2.3   SENSORS' CONDITION MONITORING WITH THE USE OF KALMAN FILTERING

The condition of the power grid's sensors can be monitored and malfunctioning of such equipment caused by grid intruders can be detected if the Kalman Filter is used to emulate the sensors' functioning. The Kalman Filter stands for a virtual (software) sensor that provides estimates of the sensors' output after using the fault-free state-space model of the power grid [191], [192], [195], [199]. As previously noted, the state-space description of the power grid is given by Eq. (5.50) while in the case of an attack to the sensors of the grid, an additive fault is assumed to affect the measurement equations. Thus the state-space model becomes

$$
\begin{aligned}
\dot{x} &= Ax + w \\
y &= Cx + v + \tilde{d}
\end{aligned}
\tag{5.52}
$$

where $\tilde{d}$ is the additive fault caused to the sensors by the network intruders. The model of the power grid is a detectable one, and this signifies that Kalman Filtering can be applied to it for state estimation purposes. After discretization, matrices A and C of the grid's state-space model are written in the form A_d and C_d. The Kalman Filter's recursion comprises the following stages:

Kalman Filter: measurement update

$$
\begin{aligned}
K(k) &= P^-(k)C_d^T[C_dP^-(k)C_d^T + R]^{-1} \\
\hat{x}(k) &= \hat{x}^-(k) + K(k)[z(k) - \hat{z}(k)] \\
P(k) &= P^-(k) - K(k)C_dP^-(k)
\end{aligned}
\tag{5.53}
$$

Kalman Filter: time update

$$
\begin{aligned}
P^-(k+1) &= A_d(k)P(k)A_d^T(k) + Q(k) \\
x^-(k+1) &= A_d\hat{x}(k)
\end{aligned}
\tag{5.54}
$$

where $P(k)$ is the covariance matrix of the state vector's tracking error, $x^-(k)$ is the value of the state vector prior to receiving measurements at the k-th sampling instant $x(k)$ is the value of the state vector after measurements $z(k)$ have been received at the k-the sampling instant, $Q(k)$ is the process noise covariance matrix and $R(k)$ is the measurement noise covariance matrix.

5.2.4 FAULT DETECTION WITH THE USE OF STATISTICAL CRITERIA

5.2.4.1 Fault detection

The residuals' sequence, is the difference between the real output of the grid sensors and the one estimated by the Kalman Filter, as shown in Fig. 5.7. This is a discrete error process e_k with dimension $m \times 1$ (here $m = 3$). Again, it is a zero-mean Gaussian white-noise process with covariance given by E_k [18], [189], [104], [239], [270], [20]. A conclusion can be stated based on a measure of certainty that the parameters of the model of the power grid sensors remain unchanged. To this end, the following *normalized error square* (NES) is defined [195]

$$
\varepsilon_k = e_k^T E_k^{-1} e_k
\tag{5.55}
$$

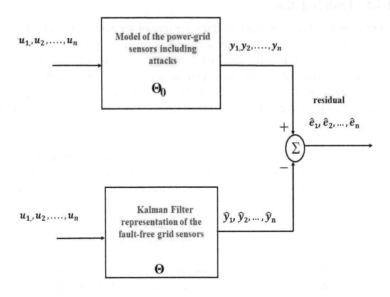

Figure 5.7 Generation of a residuals' sequence after comparing the output of the power grid's sensors against the output of the virtual sensor represented by the Kalman Filter

The normalized error square follows a χ^2 distribution. An appropriate test for the normalized error sum is to numerically show that the following condition is met within a level of confidence (according to the properties of the χ^2 distribution)

$$E\{\varepsilon_k\} = m \tag{5.56}$$

This can be achieved using statistical hypothesis testing, which is associated with confidence intervals [52], [79]. For example, a 95% confidence interval is frequently applied, which is specified using the probability region $100(1-a)$ with $a = 0.05$. Actually, a two-sided probability region is considered cutting-off two end tails of 2.5% each. For M runs the normalized error square, that is obtained from the residuals' sequence, is given by

$$\bar{\varepsilon}_k = \frac{1}{M}\sum_{i=1}^{M}\varepsilon_k(i) = \frac{1}{M}\sum_{i=1}^{M}e_k^T(i)E_k^{-1}(i)e_k(i) \tag{5.57}$$

where ε_i stands for the i-th run at time t_k. Then $M\bar{\varepsilon}_k$ will follow a χ^2 density with Mm degrees of freedom. This condition can be checked using a χ^2 test. The hypothesis holds, if the following condition is satisfied

$$\bar{\varepsilon}_k \in [\zeta_1, \zeta_2] \tag{5.58}$$

where ζ_1 and ζ_2 are derived from the tail probabilities of the χ^2 density.

5.2.4.2   Fault isolation

By applying the statistical test into the n sensors of the power system it is also possible to find out the specific sensors within the distributed power grid that has been subjected to an attack by intruders [195], [198]. In the case of a single sensor fault one has to carry out n χ^2 statistical change detection tests, where each test is applied to the i-th senor and $i = 1, 2, \cdots, n$. Actually, out of the n χ^2 statistical change detection tests, the one that exhibits the highest score identifies the sensor that has been subjected to attack.

In the case of attack to multiple sensors one can identify the subset of sensors that have been subjected to parametric change by applying the χ^2 statistical change detection test, to clusters of sensors, according to a combinatorial sequence. This means that

$$\binom{n}{k} = \frac{n!}{k!(n-k)!} \tag{5.59}$$

tests have to take place, for all sensors' clusters in the monitored power system, that finally comprise n, $n-1$, $n-2$, $\cdots$, 2, 1 sensors. Again the χ^2 tests that give the highest scores indicate sections of the grid that comprise the sensors which are most likely to have been damaged by the intruders' attack. The main requirement for applying Kalman Filtering and the statistical diagnosis test to clusters of sensors is the detectability of the associated sensors' state-space model to be preserved. More statistical fault diagnosis criteria which are based on the statistical properties of the χ^2 distribution and which ensure the detection and isolation of incipient faults and small parametric changes have been given in the *local statistical approach to fault diagnosis* [18], [270], [20].

5.2.5   USE OF THE KALMAN FILTER AS A DISTURBANCE OBSERVER

The disturbances induced to the sensors of the power grid by intruders can be identified after redesigning the Kalman Filter as a disturbance observer. The additive measurement disturbance $\tilde{d}$ and its derivatives are considered as additional state variables. Assuming that the i-th sensor is affected by the additive disturbance $\tilde{d}_i$ the following state variables appear in the extended state vector of the system: $x_7 = \tilde{d}_1$, $x_8 = \dot{\tilde{d}}_1$, $x_9 = \tilde{d}_2$, $x_{10} = \dot{\tilde{d}}_2$, $x_{11} = \tilde{d}_3$, $x_{12} = \dot{\tilde{d}}_3$. The state-space equations that describe now the evolution in time of the state variables of the power grid model are: $\dot{x}_1 = w_1$, $\dot{x}_2 = w_2$, $\dot{x}_3 = w_3$, $\dot{x}_4 = w_4$, $\dot{x}_5 = w_5$, $\dot{x}_6 = w_6$ where variables $w_i, i = 1, \cdots, 6$ denote process noise. Moreover, about the additive disturbances one has $\dot{x}_7 = x_8$, $\dot{x}_8 = \ddot{\tilde{d}}_1$, $\dot{x}_9 = x_{10}$, $\dot{x}_{10} = \ddot{\tilde{d}}_2$ and $\dot{x}_{12} = x_{12}$, $\dot{x}_{12} = \ddot{\tilde{d}}_3$.

The measurement equations of the disturbed model of the sensors become

$$y_1 = x_1 cos(\omega t) - x_2 sin(\omega t) + x_7$$
$$y_2 = x_3 cos(\omega t) - x_4 sin(\omega t) + x_9 \tag{5.60}$$
$$y_3 = x_5 cos(\omega t) - x_6 sin(\omega t) + x_{11}$$

Consequently, the state-space equations of the extended model take the f form

$$\dot{x}_e = Ax_e + w_e$$
$$y_e = C_e x_e + v_e$$

(5.61)

where $x_e = [x_1, x_2, \cdots, x_{12}]^T$ is the extended state vector, $y_e = [y_1, y_2, y_3]^T$ is the outputs vector, w_e is the vector of process noise, v_e is the vector of measurement noise, while matrices $A \in R^{12 \times 12}$ and $C \in R^{3 \times 12}$ are defined as follows

$$A_e = \begin{pmatrix} 0\,0\,0\,0\,0\,0\,0\,0\,0\,0\,0\,0 \\ 0\,0\,0\,0\,0\,0\,0\,0\,0\,0\,0\,0 \\ 0\,0\,0\,0\,0\,0\,0\,0\,0\,0\,0\,0 \\ 0\,0\,0\,0\,0\,0\,0\,0\,0\,0\,0\,0 \\ 0\,0\,0\,0\,0\,0\,0\,0\,0\,0\,0\,0 \\ 0\,0\,0\,0\,0\,0\,0\,0\,0\,0\,0\,0 \\ 0\,0\,0\,0\,0\,0\,0\,1\,0\,0\,0\,0 \\ 0\,0\,0\,0\,0\,0\,0\,0\,0\,0\,0\,0 \\ 0\,0\,0\,0\,0\,0\,0\,0\,0\,1\,0\,0 \\ 0\,0\,0\,0\,0\,0\,0\,0\,0\,0\,0\,0 \\ 0\,0\,0\,0\,0\,0\,0\,0\,0\,0\,0\,1 \\ 0\,0\,0\,0\,0\,0\,0\,0\,0\,0\,0\,0 \end{pmatrix}$$

(5.62)

$$C_e = \begin{pmatrix} cos(\omega t) & -sin(\omega t)\,0\,0\,0\,0\,1\,0\,0\,0\,0\,0 \\ 0\,0\,cos(\omega t) & -sin(\omega t)\,0\,0\,0\,0\,1\,0\,0\,0 \\ 0\,0\,0\,0\,cos(\omega t) & -sin(\omega t)\,0\,0\,0\,0\,1\,0 \end{pmatrix}$$

The disturbance observer is described by the following equation

$$\dot{\hat{x}}_e = A_e \hat{x}_e + K_e [C_e x e - C_e \hat{x}_e]$$

(5.63)

Discretization of matrices A_e and C_e gives their discrete-time equivalents A_{e_d} and C_{e_d}. Using the latter matrices one can implement the disturbance observer with the use of the Kalman Filter recursion

Kalman Filter as disturbance observer: measurement update

$$K_e(k) = P_e^-(k) C_{e_d}{}^T [C_{e_d} P_e^-(k) C_{e_d}{}^T + R_e]^{-1}$$
$$\hat{x}_e(k) = \hat{x}_e^-(k) + K_e(k)[C_e x_e(k) - C_e \hat{x}_e(k)]$$
$$P_e(k) = P_e^-(k) - K_e(k) C_{e_d} P_e^-(k)$$

(5.64)

Kalman Filter as disturbance observer: time update

$$P_e^-(k+1) = A_{e_d}(k) P_e(k) A_{e_d}^T(k) + Q_e(k)$$
$$x_e^-(k+1) = A_{e_d} \hat{x}_e(k)$$

(5.65)

The estimates of the state variables $\hat{x}_{e,7}$, $\hat{x}_{e,9}$ and $\hat{x}_{e,11}$ provide the estimates of the disturbance inputs that affect the model of the power grid sensors, that is $\tilde{d}_1$, $\tilde{d}_2$ and

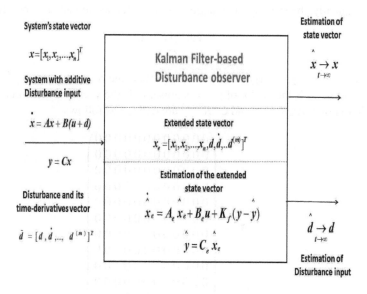

Figure 5.8   A generic diagram of the Kalman Filter-based disturbance observer

$\tilde{d}_3$, respectively. A generic diagram of the Kalman Filter-based disturbance observer for states and disturbances estimation of the power grid is given in Fig. 5.8.

5.2.6   SIMULATION TESTS

The performance of the proposed method for attack detection in the sensors of the power grid (model of a multi-area multi-machine power system) has been confirmed through simulation experiments. Since the output of the state-space model describing the sensors' functioning is $y \in R^{3 \times 1}$, the χ^2 distribution has $d = 3$ degrees of freedom. The number of samples was $M = 2000$. Thus, for a 98% confidence interval the associated lower and upper fault thresholds are $L = 2.87$ and $U = 3.13$. The first test case was concerned with the functioning of the distributed power generation system under no fault. The only external disturbance that was affecting the grid's sensors was the Gaussian measurement noise. For the case of power grid functioning that is free of attacks to its sensors, the estimation of the phasor's output provided by the Kalman Filter, and the results of the χ^2 statistical test about assessment of the grid's sensors condition are depicted in Fig. 5.9 to Fig. 5.11. In all diagrams the real values of the signals are plotted in red color while the estimated values are plotted in blue.

Moreover, the case of the power grid functioning under attacks to its sensors is depicted in Fig. 5.12 to Fig. 5.14.

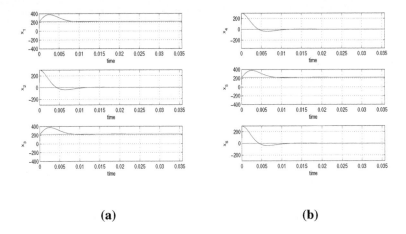

(a)   (b)

Figure 5.9   Fault-free case: (a) Estimation of state variables $x_1 - x_3$ of the power grid with the use of Kalman Filtering, (b) Estimation of state variables $x_4 - x_6$ of the power grid with the use of Kalman Filtering

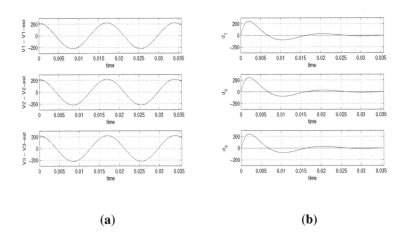

(a)   (b)

Figure 5.10   Fault-free case: (a) Estimation of voltages V_i at the measurement nodes of the power grid with the use of Kalman Filtering, (b) Estimation of the disturbance inputs that affect the sensors of the power grid after using the Kalman Filter as a disturbance observer

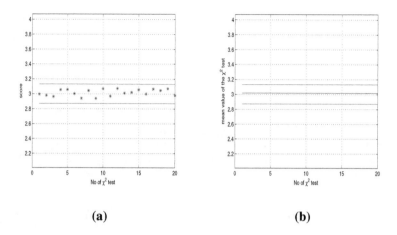

(a)   (b)

Figure 5.11 Fault-free case: (a) Consecutive χ^2 test on the Kalman Filter residuals confirm that the functioning of the grid's sensors is normal, (b) Mean value of the χ^2 test showing again the grid has not been subjected to attack

In the second test case a small additive input disturbance (due to intruders' attack) $\tilde{d} = [1.1, 0.8, 1.0]^T$ was superimposed to the voltage vector obtained from the three

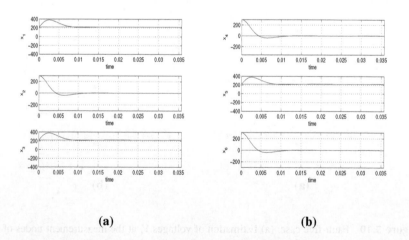

(a)   (b)

Figure 5.12 Case of grid under attack: (a) Estimation of state variables $x_1 - x_3$ of the power grid with the use of Kalman Filtering, (b) Estimation of state variables $x_4 - x_6$ of the power grid with the use of Kalman Filtering. The differences of the estimated state variables $\hat{x}_i$ from their real values x_i were practically unnoticeable

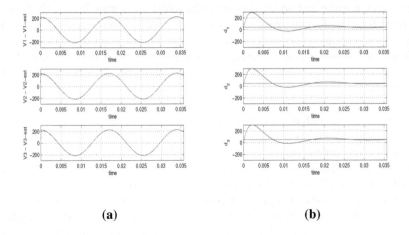

Figure 5.13   Case of grid under attack: (a) Estimation of voltages V_i at the measurement nodes of the power grid with the use of Kalman Filtering, (b) Estimation of the disturbance inputs that affect the sensors of the power grid after using the Kalman Filter as a disturbance observer

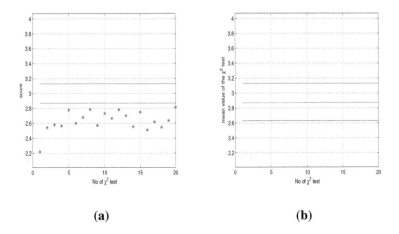

Figure 5.14   Case of grid under attack: (a) Consecutive χ^2 test on the Kalman Filter residuals confirm that the functioning of the grid's sensors is exposed to an attack, (b) Mean value of the χ^2 test showing again the grid has been subjected to attack

measurement nodes and which contained sinusoidal elements with amplitude equal to 220V. It can be distinguished that when the grid sensors were exposed to an attack the statistical χ^2 gave a clear indication about this event, even when the distortion affecting the sensor was an incipient one. By applying the statistical test to the individual sensors of the power grid one can also isolate specific sensors subjected to selective attacks.

The analyzed approach about condition monitoring of the electricity grid and detection of malicious attacks consists of three stages: (i) use of the Kalman Filter as an estimator of the state vector of the electricity grid at specific measurement points, (ii) definition of the parametric change thresholds using the properties of the χ^2 distribution, which is followed by the square of the residual of the Kalman Filter and (iii) use of the Kalman Filter as a disturbance observer so as to identify the specific waveform of the disturbance input that affects the power grid. The considered condition monitoring and attacks detection approach for the power grid is not only applicable to the electricity network shown in Fig. 5.6 but can be also applied to more types and configurations of electric power grids [90-103].

References

1. H. Abbasi Nazari, M.A. Shourehdeli, S. Simani and H.D. Bondaki, Model-based robust fault detection and isolation of an industrial gas turbine prototype using safe computing techniques, Neurocomputing, Elsevier, vol. 91, pp. 29-47, 2012.
2. A. Abdali, K. Mazlami and R. Noroozian, High-speed fault detection and location in DC microgrid system using using multi-criterion system and neural netowrk, Applied Soft Computing, Elsevier, vol. 79, pp. 341-353, 2019.
3. A. Abdali, R. Noroozian and K. Mazlami, Simultaneous control and protection schemes for DC multi-microgrids systems, Electric Power and Energy Systems, Elsevier, vol. 104, pp. 230-245, 2019.
4. H. Alavi, D. Wang and M. Luo, Short-circuit fault diagnosis for three-phase inverter sbased on voltage-space patterns. IEEE Transactions on Industrial Electronics, vol. 61, no. 10, pp. 5558-5569, 2014.
5. Y. Amirat, M.E.H.Benbouzid, E. El Ahmar, B.Bensakar and S. Tumi, A brief status on condition monitoring and fault diagnosis inwind energy conversion systems, Renewable and Sustainable Energy Reviews, vol. 13, pp. 2629-2636, 2009.
6. M. Amozegar and K. Khorasani, An ensemble of dynamic neural network identificsation for fault detection and isolation of gas turbine engine, Neural Networks, Elsevier, vol. 76, pp. 106-121, 2016.
7. O.T. An, L. Sun and L.Z. Sun, Current residual vector-based open-switch fault diagnosis of inverters in PMSM drive systems. IEEE Transactions on Power Electronics, vol. 30, no. 5, pp. 2814-2827, 2015.
8. M.R. Arahal, M.J. Duran, F. Barrero and S. Toral, Stability analysis of five-phase induction motor drives with variable third harmonic injection, Electric Power Systems Research. Elsevier, vol 80, pp. 1459-1468, 2010.
9. M.R. Arahal, C. Martin, A. Kowal, M. Castillo and F. Barrero, Cost function optimization for predictive control of a five-phase IM, Optimal Control Applications and Methods, J. Wiley, vol. 41, pp. 84-93, 2020.
10. M.R. Arahal, C. Martin, F. Barrero and M.J. Duran, Assessing variable sampling time controllers for five-phase induction motor drives, IEEE Transactions on Industrial Electronics, vol. 67, no. 4, pp. 2523-2532, 2020.
11. R.S. Arashloo, M. Salehifar, L. Romeral and V. Sala, A robust predictive current controller for healthy and open circuit faulty condition of five-phase BLDC drives applicable for wind generators and electric vehicles, Energy Conversion and Management, Elsevier, vol. 92, pp. 437-447, 2019.
12. S. Asadi, N. Vafamand, M. Moellem and T. Dragisevic, Fault reconstruction of islanded nonlinear DC microgrids: A LPV-based sliing-mode observer approach, IEEE Journal on Emerging and Selected Topics in Power Electronics, 2020.
13. M. Azanfar, J. Singh, I. Bravo-Imaz and J. Lee, Multisensor data fusion for gearbox fault diagnosis using 2D convolutional neural networks and motor current signature analysis, Mechanical Systems and Signal Processing, Elsevier, vol. 144, pp. 106681, 2020.
14. M. Babaei, J. Shi and S. Abdelwaheh, A survey on fault detection, isolation and reconfiguration methods in electric ship power systems, IEEE Access, vol. 6, pp. 9430-9441, 2018.

15. H. Badihi, Y. Zhang and H. Huang, Wind-turbine fault diagnosis and fault-tolerant torque load control against actuator faults, IEEE Transactions on Control Systems Technology, vol. 23, no.4, pp. 1351-1372, 2015.
16. S. Barsali, D. Polia, A. Pratico, R. Salvati, M. Sforna and R. Zaottini, Restoration islands supplied by gas turbines, Electric Power Systems Research, Elsevier, vol. 78, pp. 2004-2010, 2008.
17. B. Behegen and T,M. Gravdahl, Active surge control of compression system uising drive torque, Automatica, Elsevier, vol. 44, pp. 1135-1140, 2008.
18. M. Basseville and I. Nikiforov, Detection of abrupt changes: Theory and Applications. Prentice-Hall, 1993.
19. M. Basseville, A. Benveniste and Q. Zhang, Surveilliance d' installations industrielles : démarche générale et conception de l' algorithmique, IRISA Publication Interne No 1010, 1996.
20. M. Basseville, On-board component fault detection and isolation using the statistical local approach, Automatica, Elsevier, vol. 34, no. 11, pp. 1391-1415, 1998.
21. N. Bayati, A. Hajizadeh and M. Soltani, Protection in DC microgrids: a comparative review, IET Smart Grid, vol. 1, no. 3, pp. 66-75, 2018.
22. C. Beasley, X. Zhang, J. Dong, R. Brroks and G. Kumar, A survey of electric power synchrophasor network cybersecurity, ISGT Europe 2014, IEEE PES Innovative Smart Grid Technologies Conference, Oct. 2014, Istanbul, Turkey.
23. K. Bebbek, A. Merabet, M. Kesraoui, A. Tanvir and R. Beguenant, Signal-based sensor fault detection and isolation for PMSG in wind energy conversion systems, IEEE Transactions on Instrumentation and Measurement, 2017.
24. S. Belenghali, M. Benbouzid, J.F. Carpentier, T. Ahmed-Ali and J. Muntenu, Experimental validation of a marine current turbine simulator: Application on a permanent magnet synchronous generator-based system second-order sliding-mode control, IEEE Transactions on Industrial Electronics, vol. 58, no. 1, pp. 118-127, 2011.
25. S. Ben Elghali, M.Benbouzid, T. Ahmed-Ali and J.F. Charpentier, High-order sliding mode control of a marine current turbine driven doubly-fed induction generator, IEEE Journal of Oceanic Engineering, vol. 35, no. 2, pp. 402-411, 2010.
26. A. Benveniste, M. Basseville and G. Moustakides, The asymptotic local approach to change detection and model validation, IEEE Transactions on Automatic Control, vol. 32, no. 7, pp. 583-592, 1987.
27. M. Bernudes, C. Martin, I. Gonzalez-Prieto, M.J. Duran, M.R. Arahal and F. Barrero, Predictive current control in electrical drives: an illustrative review with case examples using a five-phase induction motor with drive distributed winding, IET Electric Power Applications, vol. 14, no. 8, pp. 1291-1310, 2020.
28. R. Bhargav, B.K. Bhalja and C.P. Gupta, Algorithm for fault detection and localization in a mesh-type bipolar DC microgrid network, IET Generation, Transmission and Distribution, vol. 13, no. 15, pp. 3311-3322, 2019.
29. E. Bompard, C. Guo, R. Napoli, A. Russo, M. Massera and A. Stefanini, Risk assessment of malicious attacks against power systems, IEEE Transactions on Systems, Man and Cybernetics: Part A, Systems and Humans, vol. 39, no. 5, pp. 1074-1085, 2008.
30. A. Bonfiglio, S. Cacciacarne, M. Invernizzi, R. Procopio, S. Schianno and I. Torre, Gas-turbine generating units control via feedback linearization approach, Energy, Elsevier, vol. 121, pp. 491-512, 2017.
31. Y.Q. Bostanci, M. Zafarani and B. Akin, Severity estimation of interturn short -circuit fault for PMSM, IEEE Transactions on Industrial Electronics, vol. 66, no. 9, pp. 7260-7269, 2019.

32. B. Boulkroune, M. Galvez-Garrillo and M. Kinnaert, Combined signal and model-based sensor fault diagnosis for a doubly-fed induction generator, IEEE Transactions on Control Systems Technology, vol. 21, no. 5, pp. 1771-1783, 2013.

33. T. de Bruin, K. Verbert and R. Babuska, Railway track circuit fault diagnosis using recurrent neural netwroks, IEEE Transactions on Neural Networks and Learning Systems, vol. 28, no. 3, pp. 523-533, 2017.

34. S. Budinis and N.F. Thornhill, Control of centrifugal compression via model predictive control for enhanced oil recovery applications, 2nd IFAC Workshop on Automatic Control in Offshore Oil and Gas Production, Florianapolis, Brazil, May 2015.

35. G.H. Buzan, P.R. Saalassara, W. Endo, A. Goedtel, G.F. Goday and R.H. Cunha Palacios, Electric Power Systems Research, Elsevier, vol. 143, pp. 347-356, 2017.

36. B. Cai, T. Zhao, H. Lin and M. Xin, A data-driven fault diagnosis methodology in three-phase inverters for PMSM drive systems, IEEE Transactions on Power Electronics, vol. 32, no. 7, pp. 5590-56000, 2017.

37. U. Campora, C. Cravero and R. Zaccone, Marine gas turbine monitoring and diagnostics by simulation and pattern recognition, International Journal of Naval Architecture and Ocean Engineering, Elsevier, vol. 10, pp. 617-628, 2018.

38. D.U. Campos-Delgado and D.R. Espinoza-Trejo, An observer-based diagnostic scheme for single and simultaneous open switch fault in induction motor drives, IEEE Transactions on Industrial Electronics, vol. 58, no. 2, pp. 671-679, 2011.

39. C. Cecati, F. Ciancetta and P. Siano, A multi-level inverter for photovoltaic systems with fuzzy logic control, IEEE Transactions on Industrial Electronics, vol. 57, no. 2, pp. 4115-4125, 2010.

40. C. Cecati, A.O. di Tommasso, F. Genduso, R. Miceli and G.R. Galluzzo, Comprehensive modelling and experimental testing of fault detection and management of a non-redundant fault tolerant VSI, IEEE Transactions on Industrial Electronics, vol. 69, no. 6, pp. 3946-3955, 2015.

41. A. Cecilia, S. Sahoo, T. Dragisevic, R. Costa-Castello and F. Blaabjerg, Detection and mitigation of false data in ccoperative DC microgrids with unknown constant power loads, IEEE Transactions on Power Electronics, 2021.

42. Y. Chakhchoukh and H. Ishii, Cyber attacks scenarios on the measurement function of power state estimation, 2015 American Control Conference, Chicago, USA, July 2015.

43. J. Chen, H. Li, D. Sheng and W. Li, A hybrid data-driven modelling method on sensor condition monitoring and fault diagnosis for powerplants, Electrical Power and Energy Systems, Elsevier, vol. 71, pp. 274-284, 2015.

44. H. Chen, T. Tang, N. Ait-Ahmed, M. Benbouzid, M. Machmoum and M. El-Hadi Zaim, Attraction, challenge and current status of marine current energy, IEEE Access, vol. 6, pp. 12665-12685, 2018.

45. H. Chen, Q. Li, S. Tang, N. Aid-Ahmed, J. Han, T. Wang, Z. Zhou, T. Tang and M. Benbouzid, Adaptive super-twisting ontrol of doubly salient permanent magnet generator for tidal stream turbine, International Journal of Electric Power and Energy Systems, Elsevier, vol. 128, pp. 106772-106783, 2021.

46. Z. Chen, S. Ben Elghali, M. Benbouzid, Y. Amirat, E. Elbouchikhi and G. Feld, Tidal stream turbine control: An active disturbance rejection control approach, Ocean Engineering, Elsevier, vol. 2020, pp. 107190-107198, 2020.

47. H. Chen, B. Jiang, N. Lu and W. Chen, Real-time icipient fault detection for electrical traction systems of CRH2. Neurocomputing, Elsevier, vol. 306, pp. 119-129, 2018.

48. H. Chen, B. Jiang, T. Zhang and N. Lu, Data-driven and deep learning-based detection and diagnosis of incipient faults with application to electrical traction systems, Neurocomputing, Elsevier, 2020.

49. S. Choi, B. Akin, M.R. Rahimian and H.A.Holiyar, Implementation of a fault diagnosis algorithm for induction machines based on advanced digital-signal-processing techniques, IEEE Transactions on Industrial Electronics, vol. 58, no. 1, pp. 937-948, 2011.

50. A. Chouhan, P. Gangsar, R. Porwal and C.K. Mechefske, Artificial neural network based fault diagnosis for three phase induction motors under similar operating conditions, Vibroengineering Procedia, vol. 30, pp. 55-60, 2020.

51. C. Chuang, Z. Wei, W. Zhifu and L. Ziu, The diagnosis method of stator winding faultsin PMSMs based on SOM neural networks, Energy Procedia, Elsevier, vol. 105, pp. 2295-2301, 2012.

52. J.L. Crassidis and J.L. Junkins, Optimal estimation of dynamic systems (2nd Edition), CRC Press, 2012.

53. X. Dai, T. Breiken, Z. Gao and H. Wang, Dynamic modelling and robust fault detection of a gas-turbine engine, 2008 American Control Conference, Seattle, Washington, USA, June 2008.

54. X. Dai, Z. Gao, T. Breiken and H. Wang, Disturbance attenuation in fault detection of gas turbine engines: A discrete robust observer design, IEEE Transactions on Systems, Man and Cybernetics - Part C: Applications and Reviews, vol. 39, no. 2, pp. 234-239, 2009.

55. M. Dansoko, H. NKwawo, B. Diourtê, F. Floret, R. Goma and G. Kenné, Robust multi-variable sliding mode control design for generator excitation of marine turbine in multi-machine configuration, Electrical Power and Energy Systems, Elsevier, vol. 63, pp. 423-428, 2014.

56. M. Dansoko, H. NKwawo, F. Floret, R. Goma, B. Diourté, A.Arzandé and J.C. Vannier, Marine turbine system directly connected to electrical grid: Experimental implementations using a nonlinear and robust control, Ocean Engineering, Elsevier, vol. 149, pp. 260-267, 2018.

57. M. Dansoko, H. NKwawo, B. Diourté, F. Floret, R. Goma and G. Kenné, DEcentralized sliding-mode control for marine turbines connected to grid, 11th IFAC Conference on Adaptation and Learning in Control and Signal Processing, Caen, France, July 2013.

58. P.B. Dao, W.I. Staszewski, T. Barszcz and T. Uhl, Condition monitoring and fault detection in wind turbines based on cointegration analysis of SCADA data, Renewable Energy, Elsevier, vol. 116, pp. 107-122, 2018.

59. N. Daroogheh, N. Meskin and K. Khorasani, A dual particle filter-based fault diagnosis scheme for nonlinear systems, IEEE Transactions on Control Systems Technology 2017.

60. A.M. Disqah, A. Maheri, K. Busawon and A. Kamjoo, A multivariable optimal energy strategy for standalone DC microgrids, IEEE Transactions on Power Systems, vol. 30, no. 5, pp. 2278-2287, 2015.

61. V.L. Do, L. Fillatre and I. Nikiforov, Sensitivity analysis of sequential test for detecting cyber-physical attacks, 23rd European Signal Processing Conference, Nice, France, Aug. 2015.

62. D. Dorfler, F. Pasqualetti and F. Bullo, Distributed detection of cyber-physical attacks in power networks: a waveform relaxation approach, 49th Annual Allerton Conference, Illinois, USA, Sep. 2011.

63. J. Duan, K. Zhang and L. Cheng, A novel method of fault location for single-phase microgrids, IEEE Transactions on Smart Grids, vol. 7, no. 2, pp. 915-925, 2016.

64. H. Echeikh, R. Trabelsi, A. Iqbal and M.R. Mimouni, Real-time implementation of indirect rotor flux oriented control of a five-phase induction motor with novel rotor resistance adaptation using sliding-mode observer, Journal of the Franklin Institute, Elsevier, vol. 335, pp. 2112-2141, 2018.

65. M. Echeikh, R. Trabelsi, A. Iqbal, N.Bianchi and M.F. Mimouni, Comparative study between the rotor flux oriented control and nonlinear backstepping control of a five-phase induction motor drive - an experimental validation, IET Power Electronics, vol. 9, no. 3, pp. 2510-2521, 2016.

66. M. Echeich, R. Trabelsi, H. Kesraoui, A. Iqbal and M.F. Mimouni, Torque ripple improvement of direct torque controlled five-phase induction motor drive using backstepping control, International Journal of Power Electronics and Drive Systems, vol. 11, no. 1, pp. 64-74, 2020.

67. R. Escoudero, J. Noel, J. Elizodo and J. Kirtley, Microgrid fault detection based on wavelet transformation and Park's vector approach, Electric Power Systems Research, Elsevier, vol. 152, pp. 401-410, 2017.

68. J.O. Estima and J.M. Cardoso A new approach for real-time multiple open-circuit fault diagnosis in voltage source inverters, IEEE Transactions on Industry Applications, vol. 47, no. 6, pp. 2487-2494, 2011.

69. E. Ferdjallah-Kherkhachi, E. Schaeffer, L. Loran and M. Benbouzid, Online monitoring of marime turbine insulation condition based on high frequency models: Methodology for finding the "best" indentification protocol, IEEE IECON 2014, 40th Annual Conference of the IEEE Industrial Electronics Society, Dalls, Texas, USA, Oct. 2014.

70. J. Flamband, Y. Amirat, M. Benbouzid, G. Feld and N. Ruiz, Shrouded tidal stream turbine simulation model development and experimental validation, IEEE IECON 2020, IEEE 46th Annual Conference of the Industrial Electronics Society, Singapore, Oct. 2020.

71. M. Fliess and H. Mounier, Tracking control and π-freeness of infinite dimensional linear systems, In: G. Picci and D.S. Gilliam Eds., Dynamical Systems, Control, Coding and Computer Vision, vol. 258, pp. 41-68, Birkhaüser, 1999.

72. A.J. Gallo, M.S. Turan, F. Boem, T. Parisini and G.F. Trecate, A distributed cyber-attack detection scheme with applcation to DC microgrids, IEEE Transactions on Automatic Control, vol, 65, no. 9, pp. 3800-3815, 2020.

73. Z. Gao, X. Liu and M.Z. Chen, Unknown input observer-based robust fault estimation systems corrupted by partially decoupled disturbances, IEEE Transactions on Industrial Electronics, vol. 63, no. 4, pp. 2537-2547, 2016.

74. Z. Gao, T. Brekin and H. Wang, High-gain estimator and fault tolerant design with application to a gas turbine dynamic system, IEEE Transactions on Control Systems Technology, vol. 15, no. 4, pp. 740-753, 2007.

75. Z. Gao, X. Lin and M. Chen, Unknown-input observer-based robust fault estimation for systems corrupted by partially decoupled disturbances, IEEE Transactions on Industrial Electronics, vol. 63, no. 4, pp. 2537-2547, 2016.

76. J.E. Garcia-Bracamonte, J.M. Ramirez-Cortes, J. de Jesus Rangel-Magdaleno, P. Gomez-Gil, H. Peregrino-Barreto and V. Alarcon-Aquino, An approach to MCSA-based fault detection using independentcomponents analysis and neural networks, IEEE Transactions on Instrumentation and Measurement, vol. 68, no. 5, pp.135-1361, 2019.

77. K. Gherifi, I. Garrido, A.J. Garrido, S. Bouallege and G. Haggege, Fuzzy gain scheduling of a rotational speed control for a tidal stream generator, IEEE Speedam 2018, IEEE 2018 Intl. Symposium on Power Electronics, Electrical Drives, Automation and Motion, Amalfi, Italy, June 2018.

78. B.P. Gibbs, Advanced Kalman Filtering, Least Squares and Modelling: A practical handbook, J. Wiley, 2011.

79. R.G. Gibbs, New Kalman filter and smoother consistency tests, Automatica, Elsevier, vol. 49, no. 10, pp. 3141-3144, 2013.

80. L. Gonzalez-Prieto, M.J. Duran, N. Rios-Garcia, F. Barrera and C. Martin, Open-switch fault detection in five-phase induction motor drives using model-predictive control, IEEE Transactions on Industrial Electronics, vol. 65, no. 4, pp. 3045-3055, 2018.

81. D. Gorinevsky, Fault isolation in data driven multivariate process modelling, IEEE Transactions on Control Systems Technology, vol. 23, no 5, pp. 1840-1851, 2015.

82. R. Goumouche, A. Redouane, I. El Harraki, B. Belharma and A. Hasnaoui, Optimal feedback control of nonlinear variable-speed marine current turbine using a two-mass model, Journal of Marine Science and Application, Springer, vol. 19, pp. 83-95, 2020.

83. J.T. Gravdahl, O. Egeland and S.O. Vatland, Drive torque actuation in active surge control of centrifugal compressors, Automatica, Elsevier, vol. 38, pp. 1881-1893, 2002.

84. Y. Gritli, L. Zarri, C. Rossi, F. Filippetti, G.A. Capolino and P. Casadei, Advanced diagnosis on electrical faults in wound-rotor induction machines, IEEE Transactions on Industrial Electronics, vol. 60, no. 9, pp. 4012-4024, 2013.

85. M.A. Hamida, J. de Leon and A. Glumineau, Experimental sensorless control for IPMSM by using integral backstepping strategy and adaptive high-gain observer, Control Engineering Practice, Elsevier, vol. 59, pp. 64-76, 2017.

86. X. Han, C. Liu, B. Chen and S. Zhang, Surge disturbance suppression of AMB-rotor systems in magnetically suspended centrifugal compressors, IEEE Transactions on Control Systems Technology, 2022.

87. B.P. Hand, N. Erdogan, D. Murray, P. Cronin, J. Doran and J. Murphy, Experimental testing on the influence of haft rotary lip seal misalignement for a marine hydro-kinetic turbine, Sustainable Energy Technologies and Assessments, Elsevier, vol. 50, pp. 101874-101883, 2022.

88. C.J. Harris, X. Hang and Q. Gon, Adaptive modelling, estimation and fusion from data: a neurofuzzy approach, Springer, 2002.

89. H. Hashimoto and T. Hayakawa, Distributed cyber attack detection for power network systems, IEEE CDC-ECC 2011, 50th IEEE Conference on Decision and Control and European Control Conference, Orlando, Florida, USA, Dec. 2011.

90. H. He and J. Yan, Cyber-physical attacks and defences in the smart grid: a survey, IET Cyber-Physical Systems: Theory and Applications, vol. 1, no.1, pp. 13-27, 2016.

91. A. Hooshyar and R. Irawani, Microgrids protection, Proceedings of the IEEE, vol. 105, no. 7, pp. 1332-1353, 2017.

92. K.L. del Horno, E. Segura, R. Morales and J.A. Solimonos, Exhaustive closed-loop behavior of an one degree of freedom first generation device for harvesting energy from marine currents, Applied Energy, Elsevier, vol. 276, pp. 115457-115477, 2020.

93. A. Houari, H. Renaudineau, J.P. Martin, S. Pierfederici and F. Meibody-Tabar, Flatness-based control of three-phase inverter with output LC filter, IEEE Transactions on Industrial Electronics, vol. 59, no. 7, pp. 2890-2897, 2012.

94. Q. Hu, D. Fooladivanda, Y.H. Chang and C.J. Tomlin, Secure state estimation for nonlinear power systems under cyber-attacks, 2017 American Control Conference, Seattle, USA, May 2017.

95. C.C. Hsu and C.T. Su, An adaptive forecast-based for non-Gaussian processes monitoring: with applications to equipment malfucntions detection in thermal power plants, IEEE Transactions on Control Systems Technology, vol.19, no.5, pp. 1425-1450, 2011.

96. G. Huang, E.F. Fukushima, J. She, C. Zhang and J. He, Estimation of sensor faults and unknown disturbance in current measurement circuits for PMSM drive system, Measurement, Elsevier, vol. 137, pp. 580-587, 2019.

97. T. Ince, S. Kiranyaz, L. Eren, M. Askar and M. Gabbaij, Real-time motor fault detection by 1D convolutional neural networks, IEEE Transactions on Industrial Electronics, vol. 63, no. 11, pp. 7067-7085, 2016.

98. A. Iovine, S. Benamane, G. Damm, E. De Santis and M.D. di Benedetto, Nonlinear control of a DC microgrid for the integration of photovoltaic panels, IEEE Transactions on Automation Science and Engineering, vol. 14, no. 2, pp. 524-535, 2015.

99. A. Iovine, M.J. Corrizosa, G. Damn and P. Alou, Nonlinear control for DC microgrids enabling efficient Renewable Power integration and ancillary services for AC grids, IEEE Transactions on Power Systems, vol. 34, no. 6, pp. 5136-5146, 2019.

100. A. Iovine, T. Rigaut, G. Damm, E. de Santis and M.D. de Benedetto, Power management for a DC MicroGrid integrating renewables and storages, Control Engineering Practice, Elsevier, vol. 85, pp. 59-79, 2019.

101. T. Irita and T. Namericawa, Decentralized fault detection of multiple cyber-acttacks in power network with Kalman Filters, 2015 European Control Conference, Austia, July 2015.

102. Y. Isozaki, S.Yashizawa, Y. Fujimoto, H. Ishii, I. Ono, T. Queda and Y. Hayashi, Detection of cyber-attacks against voltage control in distribution power grids with PVs, IEEE Transactions on Smart Grid, vol. 7, no. 4, pp. 1824-1835, 2015.

103. Y. Isozaki, S. Yoshikawa and Y. Fujimoto, Detection of Cyber-attacks against voltage control in distributed power grids with PVs, IEEE Transactions on Smart Grid, vol. 7, no. 4, pp. 1824-1835, 2016.

104. R. Isermann, Fault-Diagnosis Systems: An Introduction from Fault Detection to Fault Tolerance, Springer, 2006.

105. R.T. Jagaduri and G. Radman, Modeling and control of distributed generation systems including PEM fuel cell and gas turbine, Electric Power Systems Research, Elsevier, vol. 77, pp. 83–92, 2007.

106. M.J. Jahrami, A.J. Masmoud and k.J. Tseng, Design and evaluation of a new converter control strategy for near-shore tidal turbines, IEEE Transactions on Industrial Electronics, vol. 60, no. 12, pp. 5648-5659, 2021.

107. D.K. Jayamaha, N.D. Lidula and A.D. Rajapakse, Wavelet multi-resolution analysis-based ANN architecture for fault detection and localization in DC microgrids, IEEE Access, vol. 7, pp. 145371-145384, 2019.

108. H. Jeong, S. Moon and S.W. Kim, An early-stage interturn fault diagnosis of PMSMs by using negative-sequence components, IEEE Transactions on Industrial Electronics, vol. 63, no. 7, pp. 5701-5707, 2017.

109. Z. Jian, H.K. Mathews, P.G. Bonanni and S. Ruijie, Model-based sensor fault detection and isolation in gas turbine, Proc. of the 31st Chinese Control Conference, July 2012, Hefei China.

110. I. Jilasi, J.O. Estima, S.K. El-Khil, N.M. Belloaj and A.J. Marques-Cardoso, A robust observer-based method for IGSTs and current sensors fault diagnosis in voltage-source inverters of PMSM drives, IEEE Transactions on Industry Applications, vol. 53, no. 3, pp. 2894-2906, 2017.

111. F. Jurado, M. Ortega, A. Cano and J. Carpio, Neuro-fuzzy controller for gas turbine in biomass-based electric power plant, Electric Power Systems Research, vol. 60, pp. 123-135, 2002.

112. T. Kakinoki, R. Yokoyamaa, G. Fujita, K. Koyanagi, T. Funabashi, and K.Y. Lee, Shaft torque observer and excitation control for turbine–generator

torsional oscillation, Electric Power Systems Research, Elsevier, vol. 68, pp. 248-257, 2004.

113. T. Kamel, Y. Biletsky and L. Chang, Fault diagnosis for industrial grid-connected converters in the power distribution system, IEEE Transactions on Industrial Electronics, vol. 62, no. 10, pp. 6496-6505, 2015.

114. M.A. Kardan, M.H. Asemani, A. Khayutian, N. Vafamand, M.H. Khooban, T. Dragisevic and F. Blaabjerg, Improved stabilization of nonlinear DC microgrids: Cubature Kalman Filter approach, IEEE Transactions on Industry Applications, pp. 1-9, 2019.

115. A.D. Keyyon, V.M. Catterson, S.D.J. McArthur and J. Twindle, An agent-based implementation of Hidden Markov Models for Gas Turbine Condition Monitoring, IEEE Transactions on Systems, Man and Cybernetics, vol. 44, no. 2, pp. 186-195, 2014.

116. S. Khadar, H. Abu-Rub and A. Kouzou, Sensorless field-oriented control for open-end winding five-phase induction motor with parameters estimation, IEEE Open Access Journal of the Industrial Electronics Society, vol. 2, pp. 265-278, 2021.

117. K.H. Kim, Simple online fault detecting scheme for short-circuited term in a PMSM through current harmonic monitoring, IEEE Transactions on Industrial Electronics, vol. 58, pp. 2565-2569, 2011.

118. D.E. Kim, and D.C. Lee, Feedback linearization control of three-phase UPS inverter systems, IEEE Transactions on Industrial Electronics, vol. 57, no. 3, pp. 963-968, 2010.

119. E.K. Kim, F. Mwasilu, H.H. Choi and J.W. Jung, An observer-based optimal voltage control scheme for three-phase UPS Systems, IEEE Transactions on Industrial Electronics, vol. 63, no. 1, pp. 246-256, 2016.

120. C. Kloshinski, P. Ross, N. Hemdan, M. Kurrat, P. Gardinard, J. Meisner, S. Passan and A. Heinrich, Modular protection system for fault detection and selective fault clearing in DC microgrids, IET Journal of Engineering, vol. 2018, no. 12, pp. 1321-1325, 2018.

121. E. Kung, S. Day and L. Shi, The performance and limitations of ε-stealthy attacks on higher order systems, IEEE Transactions on Automatic Control, vol. 62, no. 2, pp. 941-947, 2017.

122. C. Kwan. W. Lin and I.Hwang, Sensitivity analysis for Cyber-physical systems against stealthy deception attacks, 2013 American Control Conference, ACC 2013, Washington DC USA, June 2013.

123. H.J. Laaksanen, Protection principles for future microgrids, IEEE Transactions on Power Electronics, vol. 25, no. 12, pp. 2910-2918, 2010.

124. E. Larsson, J. Aslund, E. Frisk and L. Eriksson, Health monitoring in an industrial gas turbine application by using model-based diagnosis techniques, Proc. of ASME Turbo Expo 2011, Vancouver, Canada, June 2011.

125. J.H. Lee and K.B. Lee, A fault detection method and a tolerance control in single-phase cascaded H-bridge multilevel inverter. 20th IFAC World Congress, Toulouse, France, July 2017, IFAC-PapersOnline, vol. 50, no. 1, pp. 7819-7823, 2017.

126. J. Lévine, On necessary and sufficient conditions for differential flatness, Applicable Algebra in Engineering, Communications and Computing, Springer, vol. 22, no. 1, pp. 47-90, 2011.

127. Z. Li, Y. Jiang, Q. Guo, C. Hu and Z. Peng, Multi-dimensional variational mode decomposition for bearing-crack detection in windturbines with large driving-speed variations, Renewable Energy, Elsevier, vol. 116, pp. 55-73, 2018.

128. S. Li and J. Li, Condition monitoring and diagnosis of power equipment: review and prospective, IET High Voltage, vol. 2, no. 3, pp. 82-91, 2017.

129. T. Li, R. Ma and W. Han, Virtual vector-based model predictive current control of five-phase PMSM with stator current and concentrated disturbance observer, IEEE Access, vol. 8, pp. 212635-2126456, 2020.

130. Z. Li, T. Wang, Y. Wang, T. Amirat, M. Benbouzid and D. Diallo, A wavelet threshold denoising-based imbalance fault detection method for marine current turbines, IEEE Access, vol. 8, pp. 29815-29825, 2020.

131. Q. Lin, G. Luo and J. He, Travelling-wave-based method for fault location in multi-terminal DC networks, IET Journal of Engineering, no. 13, pp. 2314-2316, 2017.

132. X. Liu and Z. Gao, Robust finite-time fault estimation for stochastic nonlinear systems with Brownian motions, Journal of the Franklin Institute, Elsevier, vol. 354, pp. 2500-2523, 2017.

133. X. Liu, Z. Gao and M.Z. Chen, Takagi-Sugeno fuzzy model based fault estimation and signal compensation with applications to wind turbines, IEEE Transactions on Industrial Electronics, vol. 64, no. 7, pp. 6678-6689, 2017.

134. F.J. Lin, T.S. Lee and C.H. Lin, Robust H_∞ controller design with recurrent neural network for linear synchronous motor drive, IEEE Transactions on Industrial Electronics, vol. 50, no. 3, pp. 456-470, 2003.

135. H. Liu, Y. Li, Y. Lin, W. Li and Y. Gu, Load reduction for two-bladed horizontal axis tidal current turbines based on individual pitch control, Ocean Engineering, Elsevier, vol. 207, pp. 107183-107192, 2020.

136. H. Liu, P. Zhang, Y. Gu, Y. Shu, J. Song, Y. Lin and W. Li, Dynamic analysis of the powertrain of 650 kW horizontal axis tidal current turbine, Renewable Energy, Elsevier, vol. 184, pp. 51-67, 2022.

137. F. Lu, H. Ju and J. Huang, An improved Extended Kalman Filter with inequality constraints for gas turbine engine health monitoring, Aerospace Science and Technology, Elsevier, vol. 58, pp. 36-47, 2016.

138. F. Lu, H. Ju and J. Huang, An improved Extended Kalman Filter with inequality constraints for gas turbine engine health monitoring, Aerospace Science and Technology, Elsevier, vol. 58, pp. 36-47, 2016.

139. F. Lu, Y. Huang, J. Huang and X. Giu, A hybrid Kalman Filtering approach based on federated framework for gas turbine health monitoring, IEEE Access, vol. 6, pp. 9841-9853, 2017.

140. Y. Lu, P. Wang, M. Jin and Y. Qi, Centrifugal compressor fault diagnosis based on qualitative simulation and thermal parameters, Mechanical Systems and Signal Processing, Elsevier, vol. 81, pp. 253-273, 2016.

141. S. Lu, G. Qian, Q. He, F. Liu, Y. Liu and Q. Wang, In situ motor fault diagnosis using enhanced nonvolutional neural network in an embedded system, IEEE Sensors Journal, 2020.

142. X. Ma, S. Zhang and K. Wang, Active surge control for magnetically suspended centrifugal compresors using a variable equilibrium point apporach, IEEE Transactions on Industrial Electronics, vol. 66, no. 12, pp. 9363-9393, 2019.

143. M.A. Mahmud, T.K. Roy, S.N. Islam, S. Saha and M.E. Haque, Nonlinear decentralized feedback linearizing controller design for islanded DC microgrids, Electric Power Components and Systems, Taylor and Francis, vol. 45, no 16, pp. 1747-1761, 2017.

144. S. Malecki, C. Bingham and Y. Zhang, Development and realization of changepoint analysis for the detection of emerging faults on Industrial Systems, IEEE Transactions on Industrial Informatics, vol. 12, no. 3, pp. 1180-1187, 2016.

145. K. Manandhar, X. Cao, F. Hu and Y. Liu, Detection of faults and attacks including false data injection attack in smart grid using Kalman Filter, IEEE Transactions on Control of Network Systems, vol. 1, no. 4, pp. 370-379, 2014.

146. K. Manandhar, X. Cao, F. Hu and Y. Liu, Combating false data injections attacks in smart grid using Kalman Filter, ICNC 2014, IEEE Intl. Conference on Computing, Networking and Communications, Hawai, Feb. 2014.

147. K. Manandhar and X. Cao, Attacks/faults detection and isolation in the Smart Grid using Kalman Filter, IEEE ICCN 2014, 23rd IEEE Intl. Conference on Computer Communicatins and Networks, Shanghai, China, Aug. 2014.

148. C. Martin, M.R. Arahal, F. Barrera and M.J. Duran, Five-phase induction motor rotor current observer for finite control set model predictive control of stator current, IEEE Transactions on Industrial Electronics, vol. 63, no. 7, pp. 4527-4538, 2016.

149. M.L. Masmoudi, E. Etien, S. Moreau and A. Sakout, Amplification of single mechanical fault signatures using full adaptive PMSM observer, IEEE Transactions on Industrial Electronics, vol. 64, no.1, pp. 615-624, 2017.

150. I. Matei, J. Baras and V. Srinivasar, Trust-based multi-agent filtering for increased smart grid security, 2012 Mediterannean Conference on Control and Automation, IEEE MED 2012, Barcelona, Spain, July 2012.

151. M.A. Mazzoletto, G.R. Bossio, C.H. De Angelo and D.R. Espinoza-Trejo, A model-based strategy for interturn short-circuit fault diagnosis in PMSM, IEEE Transactions on Industrial Electronics, vol. 64, no. 9, pp. 7218-7228, 2017.

152. L. Meegahapola and D. Flynn, Characterization of Gas Turbine Lean Blowout during frequency excursions in power networks, IEEE Transactions on Power Systems, vol. 30, no. 4, pp. 1877-1887, 2014.

153. F. Meinguet, P. Sandulescu, X. Kestelyn and E. Semail, A method for fault detection and isolation based on the processing of multiple diagnostic indices: Application to inverter faults in AC Drives. IEEE Transactions on Vehicular Technology, vol. 62, no. 3, pp. 995-1009, 2013.

154. A. Meghwani, S. Srivastava and S. Chakraborty, A non-unit protection scheme for DC microgrid based on local measurements, IEEE Transactions on Power Delivery, vol. 32, no. 1, pp. 172-181, 2016.

155. A. Meghmani, S.C. Srivestawa and S. Chakraborti. Local measurement-based techinque for estimating fault location in multi-source DC microgrids, IET Generation, Transmission and Distribution, vol. 2018, no. 13, pp. 3305-3313, 2018.

156. L.Menhour, B. d'Andréa-Novel, M. Fliess and H. Mounier, Coupled nonlinear vehicle control: Flatness-based setting with algebraic estimation techniques, Control Engineering Practice, vol. 2, pp, 135–146, 2014.

157. P. Mercorelli, A switching Kalman filter for sensorless control of a hybrid hydraulic piezo actuator using mpc for camless internal combustion engines. IEEE ICCA 2012, IEEE International Conference on Control Applications, pp. 980-985, Dubrovnil, Croatia, 2012.

158. P. Mercorelli, Parameters identification in a permanent magnet three-phase synchronous motor of a city-bus for an intelligent drive assistant, International Journal of Modelling, Identification and Control, vol. 21, no. 4, pp. 352-361, 2014.

159. S. Mochammad, Y.J. Kong, Y. Noh, S. Park and B. Ahn, Stable hybrid feature selection method for compressor fault diagnosis, IEEE Access, vol. 9, pp. 97415-97429, 2021.

160. N. Meskin, E. Naden and K. Khorasani, A multiple model-based approach for fault diagnosis of jet engines, IEEE Transactions on Control Systems Technology, vol. 21, no. 1, pp. 254-262, 2013.

161. Y. Mo, E. Garone, A. Casavola and B. Sinopoli, False data injection attacks against state estimation, in: Wireless Sensor Networks 49th IEEE Conference on Decision and Control, Dec. 2010, Arlinghton, Georgia, USA.

162. E. Mohammadi and M. Montazeri-Gh, A fuzzy-based gas turbine fault detection and identification system for full and part-load performance deterioration, Aerospace Science and Technology, vol. 46, pp. 82-93, 2015.
163. M. Monadi, C. Coch-Ciobotarou, A. Luna, J.I. Candela and R. Rodriguez, Multi-resolutional HVDC grids fault locationa ans isolation, IET Generation, Transmission and Distribution, 2016.
164. S. Moon, J. Lee, H. Jeang and S.W. Kim, Demagnetizing fault diagnosis of a PMSM-based on structure analysis of motor inductance, IEEE Transactions on Industrial Electronics, vol. 67, no. 6, pp. 3795-3803, 2016.
165. S.S. Moosavi, A. Djerdir, Y. Ait-Amira and D.A. Khaburi, ANN-based fault diagnosis of Permanent Magnet Synchronous Motor under stator winding shorted turn, Electric Power Systems Research, Elsevier, vol. 125, pp. 67-82, 2015.
166. M.M. Morato, D.J. Regner, P.R.C. Mendes, J.E. Normey-Rico and C. Bordons, Fault analysis detection and estimation for a microgrid via H_2/H_∞ LPV observers, Electrical Power and Energy Systems, Elsevier, vol. 105, pp. 823-845, 2019.
167. M. Morawiec, The adaptive backstepping control of Permanent Magnet Synchronous Motor supplied by Current Source Inverter, IEEE Transactions on Industrial Informatics, vol. 9, no. 2, pp. 1047-1055, 2013.
168. M. Morawiec and F. Wilczynski, Strong strategy of a five-phase induction machine supplied by the current source inverter with the third harmonic injection, IEEE Transactions on Power Electronics, vol. 37, no. 8, pp. 9539-9550, 2022.
169. M. Morawiec, P. Stranlowkski, A. Lewicki, J. Guzinski and F. Wilczynski, Feedback control of multiphase induction machines with backstepping techniques, IEEE Transactions on Industrial Electronics, vol. 67, no. 6, pp. 4305-4314, 2020.
170. N. Muller, S. Kouro, M. Malinovski, C.A. Rojas, M. Jasinski and. G. Estay, Medium-voltage power converter interface for multigenerator marine energy conversion system, IEEE Transactions on Industrial Electronics, vol. 64, no. 2, pp. 1061-1070, 2017.
171. M. Navi, M.R. Davoodi and N. Meskin, Sensor fault detection and isolation of an industrial gas turbine using partial kernel PCA, 9th IFAC Symposium on Fault detection, Supervision and Safety for Technical Processes, Safeprocess 2015, Paris, France, Sep. 2015.
172. Y. Nyantech, C. Edrington, S. Srivastava and D. Cortes, Application of Artificial Intelligence real-time fault detection Permanent Magnet Synchronous Machine, IEEE Transactions on Industry Applications, vol. 49, no. 3, pp. 1205-1214, 2013.
173. F. Oculi, F. Floret, M. NKwawo, R. Goma and M. Dansoko, Robust nonlinear backstepping controller for a tidal-stream generator connected to a high power grid, Journal of Energy Challenges and Mechanics, vol. 4, no. 1, 2017.
174. N. Odedele, C. Olmi and J.F. Charpentier, Power extraction strategy of a robust kW range marine tidal turbine based on a Permanent Magnet Synchronous Generator and Psssive Rectifiers, IEEE RPG 2014, IEEE 3rd Renewable Power Generation Conference, Naples, Italy, Dec. 2014.
175. E. Oland, R. Kandepu, X. Li and R.E. Ydstie, Nonlinear control of a gas turbine using backstepping, 2015 IEEE 54th Annual Conference on Decision and Control, Osaka, Japan, Dec. 2015.
176. S. Pal, B. Sidkar and J.H.Chow, Detection of malicious manipulation of synchrophasor data, 2015 IEEE Intl.Conference on Smart Grid Communications, Miami, Florida, Nov. 2015.

177. O. Palizban, K. Kauhaniemi and J.M. Guerrero, Microgrids in active network management - Part II: System Operation, Power Quality and Protection, Renewable and Sustainable Energy Reviews, vol. 36, pp. 440-451, 2014.

178. S. Pan, T. Mervis and U. Adhikari, Classification of disturbances and cyber-attacks in power systems with heterogeneous time-synchronization data, IEEE Transactions on Industrial Informatics, vol. 11, no. 3, pp. 650-662, 2015.

179. J.D. Park, J. Candelaria, L. Ma and K. Duan, DC Ring-Bus microgrid fault protection and identification of fault location, IEEE Transactions on Power Delivery, vol. 28, no. 4, pp. 2574-2584, 2013.

180. F. Pasqualetti, F. Dorfler and F. Bullo, Attack detection and identification in Cyber-physical systems, IEEE Transactions on Automatic Control, vol. 58, no. 11, pp. 2715-2729, 2013.

181. T. Peng, S. Zhou, Z. Lin, Y. Sun and Y. Man, Fault diagnosis for shunt hybrid active power filters with open-circuit fault based on voltage distortion, IEEE CAC 2015, Chinese Automation Congress 2015, Wuhan, China, 2015.

182. Z. Peng, G. Liu, D. Zhou, F. Hsu and D. Sun, Two-channel false data injection attacks against output tracking control of networked systems, IEEE Transactions on Industrial Electronics, vol. 63, no. 5, pp. 3242-3251, 2016.

183. H.T. Pham, J.M. Bourgeot and M. Benbouzid, Fault tolerant finite control set-model predictive control for marine current turbine applications, IET Renewable Power Generation, vol. 12, no. 4, pp. 415-421, 2018.

184. B. Pourbobaee, N. Meskin and K. Khorasani, Sensor fault detection, isolation and identification using multiple model-based hybrid Kalman Filter for Gas turbine engines, IEEE Transactions of Control Systems Technology, vol. 26, no. 4, pp.1184-1200, 2016.

185. M. Priestley, J.E. Fletcher and C. Tan, Space-vector PWM technique for five-phase open-end winding PMSM drive operating in the overmodulation region, IEEE Transactions on Industrial Electronics, vol. 65, no. 9, pp. 6816-6827, 2018.

186. S. Rahme and N. Meskin, Adaptive sliding-mode observer for sensor fault diagnosis of an industrial gas turbine, Control Engineering Practice, Elsevier, vol. 98, pp. 57-74, 2015.

187. G.G. Rigatos and S.G. Tzafestas, Neural structures using the eigenstates of the Quantum Harmonic Oscillator, Open Systems and Information Dynamics, Springer, vol. 13, no. 1, 2006.

188. G.G. Rigatos and S.G. Tzafestas, Extended Kalman Filtering for Fuzzy Modelling and Multi-Sensor Fusion. Mathematical and Computer Modelling of Dynamical Systems, Taylor & Francis, vol. 13, pp. 251-266, 2007.

189. G. Rigatos and Q. Zhang, Fuzzy model validation using the local statistical approach, Fuzzy Sets and Systems, Elsevier, vol. 60, no. 7, pp. 882-904, 2009.

190. G.G. Rigatos, P. Siano and A. Piccolo, A neural network-based approach for early detection of cascading events in electric power systems, IET Journal on Generation Transmission and Distribution, vol. 3, no. 7, pp. 650-665, 2009.

191. G. Rigatos, Modelling and control for intelligent industrial systems: adaptive algorithms in robotcs and industrial engineering, Springer, 2011.

192. G. Rigatos, Advanced models of neural networks: nonlinear dynamics and stochasticity in biological neurons, Springer, 2013.

193. G. Rigatos, P. Siano and N. Zervos, Sensorless Control of Distributed Power Generators With the Derivative-Free Nonlinear Kalman Filter, IEEE Transactions on Industrial Electronics, vol. 61, no. 11, pp. 6369-6382, 2014.

194. G. Rigatos, P. Siano, N. Zervos and C. Cecati, Control and disturbances compensation for doubly Fed induction generators using the derivative-free nonlinear Kalman Filter, IEEE Transactions on Power Electronics, vol. 30, no. 10, pp. 5532-5547, Oct. 2015.

195. G. Rigatos, Nonlinear control and filtering using differential flatness theory: applications to electromechanical systems, Springer, 2015.
196. G. Rigatos, P. Siano and C. Cecati, A new non-linear H-infinity feedback control approach for three-phase voltage source converters, Electric Power Components and Systems, Taylor and Francis, vol. 44, no. 3, pp. 302-312, 2015.
197. G. Rigatos, P. Siano, P. Wira and F. Profumo, Nonlinear H-infinity feedback control for asynchronous motors of electric trains, Journal of Intelligent Industrial Systems, Springer, vol. 1, no. 2, pp. 85-98, 2015.
198. G. Rigatos, Intelligent Renewable Energy Systems: Modelling and Control, Springer, 2016.
199. G. Rigatos, State-space appproaches for modelling and control in financial engineering: systems theory and machine learning methods, Springer, 2017.
200. G. Rigatos, P. Siano, F. Marignetti and I. Gros, A nonlinear optimal control aporoach for PM Linear Synchronous Motors, IEEE INDIN 2018, IEEE 16th Intl. Conference on Industrial Informatics, Porto, Portugal, July 2018.
201. G. Rigatos, N. Zervos, D. Serpanos, V. Siadimas, P. Siano and M. Abbaszadeh, Condition monitoring of gas-turbine power units using the Derivative-free nonlinear Kalman Filter, IEEE SEST 2018, IEEE 2018 Intl. Conference on Smart Energy Systems and Technologies, Valencia, Spain, June 2018.
202. G. Rigatos and K. Busawon, Robotic manipulators and vehicles: Control, estimation and filtering, Springer, 2018.
203. G. Rigatos, N. Zervos, P. Siano and M. Abbaszadeh, Nonlinear optimal control for DC industrial microgrids, Cyber-Physical Systems, Taylor and Francis, vol. 5, no. 4, pp. 231-253, 2019.
204. G. Rigatos and E. Karapanou, Advances in applied nonlinear optimal control, Cambridge Scholars Publications, 2020.
205. G. Rigatos, N. Zervos, P. Siano, M. Abbaszadej and M. Hamida, Nonlinear optimal control for the synchronization of distributed marine-turbine power generation units, Electric Power Components and Systems, Taylor and Francis, vol. 49, no. 4-5, pp. 436-457, 2021.
206. G. Rigatos, M. Abbaszadeh and P. Siano, Control and estimation for dynamical nonlinear and partial differential equation systems: Theory and applications, IET Publications, 2022.
207. G. Rigatos, M.A. Hamida, M. Abbaszadeh and P. Siano, A nonlinear optimal control approach for shipboard AC/DC microgrids, Electric Power Systems Research, Elsevier, vol. 209, pp. 108024, 2022.
208. J.A. Riveros, F. Barrero, E. Levi, M.J. Duran, S. Toral and M. Jones, Variable-speed five-phase induction motor drive based on Predictive Torque Control, IEEE Transactions on Industrial Electronics, vol. 60, no. 8, pp. 2957-2968, 2018.
209. M.A. Rodriquez-Blanco, A. Vasquez-Perez, L. Henrnandez-Gonzalez, V. Golikov, J. Aguayar-Alquiera and M. May-Alarcan, Fault detection for IGBT using adaptive threshold during turn-on transient, IEEE Transactions on Industrial Electronics, vol. 62, no. 3, pp. 1975-1984, 2016.
210. J.A. Rosero, L. Romeral, J.A. Ortega and E. Rosero, Short-circuit detection by means of empirical mode decomposition and Wigner-Ville distribution for PMSM running under dynamic condition, IEEE Transactions on Industrial Electronics, vol. 56, no. 11, pp. 4534-4547, 2009.
211. L. Rothenhagen and F.W. Fuchs, Doubly-fed induction generator model-based sensor fault detection and control loop reconfiguration, IEEE Transactions on Industrial Electronics, vol. 56, no. 10, pp. 4229-4238, 2009.
212. K. Saad, K. Abdellah, H. Ahmed and A. Iqbal, Investigation of SVM-backstepping sensorless control of five-phase open-end winding induction

motor based on model reference adaptive system and parameter estimation, Engineering Science and Technology, Elsevier, vol. 22, pp. 1013-1026, 2019.

213. H. Saavedra, J.C. Urresty, J.R. Ribo and L. Romeral, Detection of interturn faults in PMSMs with different winding configurations, Energy Conversion and Management, Elsevier, vol. 79, pp. 534-542, 2014.

214. Z.N. Sadaghi Vanini, K. Khorasani and N. Meshkin, Fault detection and isolation of a dual speed gas turbine engine dynamic neural networks and multiple model approach, Information Sciences, Elsevier, vol. 259, pp. 234-251, 2014.

215. S. Saha, T.K. Roy, M.A. Mahmud, M.E. Haque and S.N. Islam, Sensor-fault and cyber-attack resilient operation of DC microgrid, Electrical Power and Energy Systems, Elsevier, vol. 93, pp. 540-554, 2018.

216. A. Salar, N. Meskin and K. Khorasani, A novel affine qLPV Model Derivation Method for fault diagnosis H_∞ performance improvement,ACC 2015, IEEE 2015 American Control Conference, Chicago, USA, July 2015.

217. M. Salehifar, M. Moreno-Eguilaz, G. Putrus and P. Barras, Simplified fault tolearant finite control set model predictive controol of a five-phase inverter supplying BLDC moor in electric vehicle drive, Electric Power Systems Research, Elsevier, vol. 132, pp. 56-66, 2016.

218. M. Sanchez-Para and C.Verde, Analytical redundancy for a gas turbine of a combined-cycle power plant, Proc. of the 2006 American Control Conference, Minneapolis, Minnesota, USA, June 2006.

219. H. Sanchez, T. Escabet, V. Puig and P.F. Odgaard, Fault diagnosis of an advanced wind-turbine benchmark using interval-based ARRs and observers, IEEE Transactions on Industrial Electronics, vol. 62, no. 6, pp. 3783-3793, 2015.

220. A. Sarikhani and O.A. Mohammed, Inter-turn fault detection in PM synchronous machines, by physics-based back electromotive force estimation, IEEE Transactions on Industrial Electronics, vol. 60, no. 8, pp. 3472-3484, 2013.

221. M. Schimmack, B. Haus and P. Mercorelli, An extended Kalman filter as an observer in a control structure for health monitoring of a metal–polymer hybrid soft actuator, IEEE/ASME Transactions on Mechatronics, vol. 23, no. 3, pp. 1477-1487, 2018.

222. C. Shu, C.Y. Ting, Y.T. Jian and W. Xun, A novel diagnostic technique for open-circuited faults of inverters based on output line-to-line voltage model, IEEE Transactions on Industrial Electronics, vol. 68, no. 2, pp. 4412-4421, 2016.

223. S. Simani, C. Fantuzzi and S. Beghello, Diagnosis techniques for sensor faults of industrial processes, IEEE Transactions on Control Systems Technology, vol. 8, no. 3, pp. 848-855, 2000.

224. S. Simani, R.J. Patton, S. Doley and A. Pike, Identification and fault diagnosis of an industrial gas-turbine prototype model, Proc. of the 39th IEEE Conference on Decision and Control, Sydney, Australia, Dec. 2000.

225. S. Simani and R.J. Patton, Model-based fault diagnosis approaches with application to an industrial gas turbine simulator, Proc. of the European Control Conference, Budapest, Hungary, Aug. 2009.

226. D. Simon, A game theory approach to constrained minimax state estimation, IEEE Transactions on Signal Processing, vol. 54, no. 2, pp. 405-412, 2006.

227. S. Sina Tayarani-Bathaie, Z.N. Sadaghi Vanini and K. Khorasani, Dynamic neural network-based fault diagnosis of gas-turbine engines, Neurocomputing, Elsevier, vol. 125, pp. 153-165, 2014.

228. S. Sina Tayarani-Bathaie and K. Khorasani, Fault detection and isolation of gas turbine engines using a bank of neural networks, Journal of Process Control, Elsevier, vol. 36, pp. 22-41, 2015.

229. Y. Mo and B. Sinopoli, False data injection attacks in control systems, 1st Workshop of Secure Control Systems, Stockholm, Sweden, 2010.
230. H. Sira-Ramirez and S. Agrawal, Differentially Flat Systems, Marcel Dekker, New York, 2004.
231. R. Soleymani, M.A. Nekani and M. Mourefianpour, A novel robust fault detection method for induction motor rotor by using unknown input observer, Systems Science and Control Engineering, Taylor and Francis, vol. 7, no. 1, pp. 109-115, 2019.
232. A. Souza Guedes, S. Magalhues Silva, B. de Jesus Cardoso Filho and C. A. Conceicao, Evaluation of electrical insulation in three-phase induction motors and classification of failures using neural networks, Electric Power Systems Research, Elsevier, vol. 140, pp. 263-273, 2016.
233. S. Tan, P. Xie, JM. Guerrero, J.C Vasquez and R. Han, Cyber-attack detection for converter-based distributed DC microgrids: Observer-based approaches, IEEE Industrial Electronics Magazine, 2021.
234. Y. Tang, Y. Zhang, A. Hasankhani and J. Van Zwieten, Adaptive super-twisting sliding mode control for ocean curren turbine-driven permanent magnet synchronous generator, IEEE ACC 2020, IEEE 2020 American Control Conference, Denver, Colorado, USA, July 2020.
235. A. Teixeira, H. Sandberg and K.H. Johanson, Networked control systems under cyber attacks with applications to power networks, ACC 2010, IEEE American Control Conference, Baltimore, USA, June 2010.
236. A. Teixeira, I. Shames, H. Snadberg and K. Johanson, Distributed fault detection and isolation resilient to network model uncertainties, IEEE Transactions on Cybernetics, vol. 14, no. 11, pp. 2014-2037, 2014.
237. R. Teixeira-Pinto, P. Bauer, S.F. Rodrigues, E.J. Wiggelinkhuizer, J. Picrik and B. Ferreira, A novel distributed direct-voltage control strategy for grid integration of offshore wind energy systems through MTDC network, IEEE Transactions on Industrial Electronics, vol. 60, no. 6, pp. 2429-2441, 2013.
238. Y. Teng, D. Hu, F. Wu, R. Zhang and F. Gao, Fast economic model predictive control for marine current turbine generator system, Renewable Energy, Elsevier, vol. 166, pp. 108-116, 2020.
239. N. Tleis, Power system modelling and fault analysis: Theory and practice, Elsevier, 2008.
240. G. Torrisi, V Jaramillo, J.R. Ottewill, M. Mariethoz, M. Morari and R.S. Smith, Active surge control of electrically driven cntrifugal compressors, 2015 European Control Conference, Linz, Austria, July 2015.
241. G. Torrissi, S. Grammatico, A. Cortinovis, M. Mercangoz, M. Morari and R.S. Smith, Model predictive approaches for active surge control in centrifugal compressors, IEEE Transactions on Control Systems Technology, vol. 29, no 6, pp. 1947-1960, 2019.
242. S. Toumi, S. Benelghali, M. Trabelsi, E. Elbouchikhi, Y. Amirat, M. Benbouzid and M.F. Mimouni, Modelling and simulation of a PMSG-based marine current turbine system under faulty rectifier conditions, Electric Power Components and Systems, Taylor and Francis, vol. 45, no. 7, pp. 715-725, 2017.
243. G.J. Toussaint, T. Basar and F. Bullo, H_∞ optimal tracking control techniques for nonlinear underactuated systems, in Proc. IEEE CDC 2000, 39th IEEE Conference on Decision and Control, Sydney, Australia, Dec. 2000.
244. M. Tribelsi and E. Semail, Virtual current vector-based method for inverter open-switch and open-phase fault diagnosis in multi-phase permanent magnet synchronous motor drives, IET Electric Power Applications, pp. 1-16, 2021.
245. J. Villagra, B. d'Andrea-Novel, H. Mounier and M. Pengov, Flatness-based vehicle steering control strategy with SDRE feedback gains tuned via a

sensitivity approach, IEEE Transactions on Control Systems Technology, vol. 15, pp. 554-565, 2007.

246. B. Whitby and C.E. Ugalde-Loo, Performance of pitch and stall regulated tidal stream turbines, IEEE Transactions on Sustainable Energy, vol. 5, no. 1, pp. 64-73, 2014.

247. L.X. Wang, A course in fuzzy systems and control, Prentice-Hall, 1998.

248. Y. Wang and X. Yu, New coordinated control design for thermal power generation units, IEEE Transactions on Industrial Electronics, vol. 57, no. 1, pp. 3848-3856, 2010.

249. T. Wang, L. Liu, J. Zhang, E. Scaeffer and Y. Wang, A M-EKF fault detection strategy of insulation system for marine current turbine, Mechanical Systems and Signal Processing, Elsevier, vol. 115, pp. 269-280, 2019.

250. T. Wang, L. Liang, S.K. Gurumurthy, F. Pponci, A. Monti, Z. Yang and RW. DeDoncker, Model-based fault detection and isolationin DC microgrids using optimal observers, IEEE Journal on Emerging and Selected Topics in Power Electronics, 2020.

251. H. Wang, A. Meng, Y. Liu, X. Fu and G. Cao, Unscented Kalman Filter-based internal state estimation of cyber-physical energy system for detection of dynamic network, Energy, Elsevier, vol. 188, pp. 116076, 2019.

252. P. Wang, B. Zhao, H. Chang, B. Huang, W. He, Q. Zhang and F. Zhu, Study of the performance of a 300 kW counter rotating type horizontal axis tidal turbine, Ocean Engineering, Elsevier, vol. 255, pp. 111446-111456, 2022.

253. J. Wei, T. Xie and T. Wang, A VMD denoising-based imbalance fault detection method for marine current turbine, IEEE IECON 2020, 46th Annual Conference of the IEEE Industrial Electronics Society, Singapore, Oct. 2020.

254. J.H. Won and S.M. Seong, EKF-based fault detection and isolation for FET of induction motor inverter. 2015 15th Intl. Conference on Control, Automation and Systems, IEEE ICCAS 2015, Oct. 2015, Busan Korea, 2015.

255. C. Wu, C. Guo, Z. Xie, F. Ni and H. Liu, A signal-based fault detection and tolerance control method of current sensor for PMSM drive, IEEE Transactions on Industrial Electronics, vol. 65, no. 12, pp. 9646-9657, 2018.

256. T. Xi, Z. Li, T. Wang, M. Shi and Y. Wang, An integration fault detection method using stator voltage for marine current turbines, Ocean Engineering, Elsevier, vol. 226, pp. 108808-108818, 2021.

257. S. Xiang and J. Li, Cascade model predictive current control for five-phase permanent magnet synchronous motor, IEEE Access, vol. 10, pp. 88812-88820, 2022.

258. T. Xie, T. Wang and D. Diallo, Marine current turbine imbalance fault detecdtion method based on angular resampling, 21st IFAC World Congress, Berlin, Germany, July 2020.

259. T. Xie, T. Wang, Q. He, D. Diallo and C. Claramunt, A review of current issues of marine current turbine fault detection, Ocean Engineering, Elsevier, vol. 2018, pp. 108194-108209, 2020.

260. T. Xie, Z. Li, T. Wang, M. Shi and Y. Wang, An integration fault detection method using stator voltage for marine current turbines, Ocean Engineering, Elsevier, vol. 226, pp. 108808-108818, 2021.

261. C. Xiong, H. Xu, T. Guan and P. Zhou, A constraint switching frequency multiple-vector-based model predictive current control of five-phase PMSM with nonsinusoidal back EMF, IEEE Transactions on Industrial Electronics, vol. 67, no. 3, pp. 1695-1707, 2020.

262. W. Yang, K.Y. Lee, T.S. Yunker and H. Ghezel-Ayagh, Fuzzy fault diagnosis and accomodation system for hybrid fuel-cell / Gas-turbine power plant, IEEE Transactions on Energy Conversion, vol. 15, no. 4, pp. 1187-1194, 2010.

263. S.K. Yee, J.V. Milanovic and F.M. Hughes, Phase compensated gas turbine governor for damping oscillatory modes, Electric Power Systems Research, Elsevier, vol. 79, pp. 1192-1199, 2009.

264. Y. Yu, Y. Zhou, X. Huang and D. Xu, Current sensor fault diagnosis and tolerant control for VSI-based induction motor drives, IEEE Transactions on Power Electronics, vol. 33, no. 5, pp. 4238-4248, 2018.

265. M.A. Zai, R. Retan, A.R. Mills and R.E. Harrison, Prognostics of gas-turbine engine: An integrated approach, Expert Systems with Applications, Elsevier, vol. 42, pp. 8472-8483, 2015.

266. J. Zarei and E. Shokri, Robust sensor fault detection based on nonlinear unknown input observer, Measurement, Elsevier, vol. 48, pp. 353-367, 2014.

267. Q. Zhang and A. Benveniste, Wavelet networks, IEEE Transactions on Neural Networks, vol. 3, no. 6, pp. 869-898, 1993.

268. Q. Zhang, M. Basseville and A. Benveniste, Early warning of slight changes in systems, Special Issue on Statistical Methods in Signal Processing and Control, Automatica, Elsevier, vol. 30, no.1, pp. 95-113, 1994.

269. Q. Zhang, Fault detection and isolation with nonlinear black-box models, Proc.SYSID 1997, Kitakyushu, Japan, 1997.

270. Q. Zhang, M. Basseville and A. Benveniste, Fault detection and isolation in nonlinear dynamic systems : A combined input-output and local approach, Automatica, Elsevier, vol. 34, no. 11, pp. 1359-1373, 1998.

271. Q. Zhang, Adaptive observer for multiple-input-multiple-output (MIMO) linear time-varying systems, IEEE Transactions on Automatic Control, vol. 47, no. 3, pp. 525-529, 2002.

272. Q. Zhang and M. Basseville, Statistical detection and isolation of additive faults in linear time-varying systems, Automatica, Elsevier, vol. 50, no. 10, pp. 2527-2538, 2014.

273. Q. Zhang, Stochastic Hybrid System Actuator Fault Diagnosis by Adaptive Estimation. IFAC Safeprocess 2015. 9th IFAC Symposium on Fault Detection, Supervision and Safety for Technical Processes, Paris, France, 2015.

274. Q. Zhang, Adaptive Kalman Filter for actuator fault diagnosis. IFAC WC 2017, 20th IFAC World Congress, 2017.

275. Q. Zhang, Adaptive Kalman Filter for actuator fault diagnosis, Automatica, Elsevier, vol. 93, pp. 333-342, 2018.

276. K. Zhang, B. Jiang, X.G. Yan and Z. Man, Sliding-mode observer-based incipient sensor fault detection and application to high-speed railway traction devices, ISA Transactions, Elsevier, vol. 63, pp. 49-59, 2016.

277. Y. Zhang, Y. Fan and W. Du, Nonlinear process monitoring using regression and reconstruction method, IEEE Transactions on Automation Science and Engineering, vol. 13, no. 3, pp. 1343-1354, 2016.

278. J. Zhang, J. Zhou, D. Zhou and C. Huang, High-performance fault diagnosis in PWM voltage-source inverters for vector controlled induction motor drives, IEEE Transactions on Power Electronics, vol. 39, no. 11, pp. 6087-6099, 2014.

279. M. Zhang, T. Wang, T. Tang, M. Benbouzid and D. Diallo, Imbalance fault detection of marine current turbbine under condition of wave and turbulence, IEEE IECON 2016, 42nd Annual of the IEEE Industrial Electronics Society, Florence, Italy, Oct, 2016.

280. M. Zhang, T. Wang and T. Tang, Multi-doamin reference method for fault detection of marine current turbine, IEEE IECON 2017, 43rd Annual Conference of the IEEE Industrial Electronics Society, Beijing, China, Dec. 2017.

281. M. Zhang, T. Wang, T. Tang, M. Benbouzid and D. Diallo, An imbalance fault detection method based on data normalization and EMD for marine current turbines, ISA Transactions, Elsevier, vol. 68, pp. 302-312, 2017.

282. R. Zhang, H. Tao, L.Wu and Y. Guan, Transfer learning with neural networs for bearing fault diagnosis in changing working conditions, IEEE Access, vol. 5, pp. 14347-14357, 2017.

283. K. Zhang, B. Jiang, X.G. Yan and Z. Mao, Incipient voltage sensor fault isolation for rectifier in railway electric traction system, IEEE Transactions on Industrial Electronics, vol. 64, no. 8, pp. 6763-6774, 2017.

284. Q. Zhang, F. Giri and T. Ahmed-Ali, Regularized Adaptive Observer to Address Deficient Excitation. IFAC ALCOS 2019, 13th IFAC Workshop on Adaptive and Learning Control Systems, Winchester, UK, 2019.

285. Y. Zhang, E. Fernandez-Rodriguez, J. Zhang, Y. Zheng, J. Zhang, H. Gu, W. Zang and X. Lin, A review on numerical development of tidal stream turbine performance and make prediction, IEEE Access, vol. 8, pp. 79325-79337, 2020.

286. Z. Zhou, F. Scuiller, J.F. Carpentier, M. Benbouzid and T. Tong, Power limitation control for a PMSG-based marine-current turbine at high tidal speed and strong sea state, IEEE 2019 Conference on Electric Machines and Drives, Chicago, USA, May 2013.

287. Z. Zhou, S. Ben Elghali, M. Benbouzid, Y. Amirat, E. Ilbouchikhi and G. Feld, Control strategies for tidal stream turbine system: a comparative study of ADRC, PI and High-order sliding-mode controls, IEEE IECON 2019, IEEE 45th Intl. Conference of the Industrial Electronics Society, Lisbon, Portugal, Oct. 2019.

288. Y. Zhou, Z. Yan, Q. Duan, L. Wang and X. Wu, Direct torque control strategy of five-phase PMSM with load capacity enhancement, IET Power Electronics, vol. 12, no. 3, pp. 598-606, 2019.

289. X. Zhou, T. Wang, M. Zhang, T. Xie, Z. Li and S. Pandey, A control strategy for active disturbance rejection control based on marine current turbine, IEEE IECON 2019, IEEE 45th Annual Conference of the Industrial Electronics Society, Lisbon, Portugal, Oct. 2019.

290. X. Zhou, T. Wang and D. Diallo, An active disturbance-rejection sensorless control strategy based on sliding-mode observer for marine current turbine, ISA Transactions, Elsevier, 2021.

291. T. Zwerger and P. Mercorelli, Combining SMC and MTPA Using an EKF to Estimate Parameters and States of an Interior PMSM, IEEE ICC 2019, 20th International Carpathian Control Conference, pp. 1-6, Krakow-Wieliczka, Poland, 2018.

292. T. Zwerger and P. Mercorelli, A Dual Kalman Filter to Identify Parameters of a Permanent Magnet Synchronous Motor, IEEE ICSTCC 2020, 24th International Conference on System Theory, Control and Computing, Sinaia, Romania, 2020.

Index

B

Backstepping-type control, 39

C

Canonical (Brunovsky) form, 84
   flatness-based control with
      transformation into, 67, 68–81
Central Limit Theorem (CLT), 35
χ^2 (chi-square) distribution, 10, 87, 88,
   93, 99, 108, 114
χ^2 (chi-square) test, 11, 12, 35–36, 93,
   124
   statistical change detection, 87–88
Condition monitoring methods, 83
Cyberattacks
   affecting DC microgrids,
      101–102
   affecting gas-turbine power
      generation units, 145–146
   affecting steam turbine power
      generation units, 172, 173

D

DC microgrids, electric power
   dynamic model, 100–102
   fault detection, 106–107
   fault diagnosis, 98–113
      with H-infinity Kalman Filter,
         104–106
   fault isolation, 107
   faults and cyberattacks affecting,
      101–102
   simulation tests, 108–113
   state-space model, 99
Derivative-free nonlinear Kalman Filter,
   86, 140, 178
   state estimation using, 153–154
Diffeomorphisms, 38
Differential flatness
   marine turbine power generation
      system, 84–85

properties
   gas-turbine power generation
      units, 146–148
   steam turbine power generation
      units, 172–177
   wind power generation systems,
      196–197
dq reference frame, 4
Dynamic model
   DC microgrids, 100–102
   gas compressors actuated by
      five-phase induction motor,
      114–118
   gas-turbine power generation units,
      140–146
   marine turbine power generation
      system, 39, 40–42
   Permanent Magnet Linear
      Synchronous Machine/Motor,
      19–21
   steam turbine power generation
      units, 169–172
   three-phase inverters,
      2–6
   wind power generation systems,
      192–195

E

Electrical part of PMLSM,
   20–21
Electricity grid, *see* power grid
Electric power DC microgrids
   dynamic model, 100–102
   fault detection, 106–107
   fault diagnosis for, 98–113
      with H-infinity Kalman Filter,
         104–106
   fault isolation, 107
   faults and cyberattacks affecting,
      101–102
   simulation tests, 108–113

state-space model, 99
  approximate linearization of,
    102–104
Extended Kalman Filter (EKF), 16

F
Fault detection, 10–11, 26–29
  electric power DC microgrids,
    106–107
  gas compressors actuated by
    five-phase induction motor,
    123–124
  gas-turbine power generation units,
    155–156
  power grid, 206–207
  statistical tests for, 86–87
  steam turbine power generation
    units, 180–181
  wind power generation systems,
    200–201
Fault diagnosis
  electric power DC microgrids,
    98–113
  H-infinity Kalman Filter,
    104–106
  marine turbine power generation
    system, 82–97
  fault detection, statistical tests
    for, 86–87
  fault isolation, statistical tests
    for, 87–88
  Kalman Filter-based disturbance
    observer, 85–86
  simulation tests, 88–97
  model-based, 1–16
  model-free, 16–36
  for steam turbine power generation
    units, 167–190
Fault isolation, 11–12, 29–30
  electric power DC microgrids, 107
  gas-turbine power generation units,
    156–157
  power grid, 207
  statistical tests for, 87–88

steam turbine power generation
    units, 181–182
  wind power generation systems,
    200–201
Faults
  affecting DC microgrids, 101–102
  affecting gas-turbine power
    generation units, 145–146
  affecting steam turbine power
    generation units, 172, 173
Feedback control law, 45, 176
Feed-forward neural networks (FNN),
    21–23
First-order Taylor series expansion, 38,
    113, 118
Five-phase induction motor
  gas compressors actuated by
    approximate linearization of,
      118–122
    dynamic model, 114–118
    fault detection, 123–124
    fault diagnosis for, 113–138
    fault isolation, 124–126
    simulation tests, 126–138
    state estimation with H-infinity
      Kalman Filter, 122–123
Flatness-based control
  for marine turbine power
    generation system, 50–56, 67,
      68–81
  in successive loops, 53–56, 70–72,
      82
Fourier series, 22
Fuzzy multi-model-based control
  for marine turbine power
    generation system, 75–77, 82

G
Gas compressors
  actuated by five-phase induction
    motor
    approximate linearization of,
      118–122
    dynamic model, 114–118

fault detection, 123–124
fault diagnosis for, 113–138
fault isolation, 124–126
simulation tests, 126–138
state estimation with H-infinity
Kalman Filter, 122–123
Gas-turbine power generation units,
139–140
differential flatness-properties of,
146–148
dynamic model, 140–146
fault detection, 155–156
fault isolation, 156–157
faults and cyberattacks affecting,
145–146
simulation tests, 157–167
state estimation using
derivative-free nonlinear
Kalman Filter, 153–154
state-space description of, 140–145
state-space model, 146
statistical fault diagnosis using
Kalman Filter, 155–157
transformation, into input-output
linearized form, 148–153
Gauss-Hermite neural networks, 19, 30
Gauss-Hermite series expansion, 23–25

H

H-bridge (transistors) circuit, 6
H-infinity control, 38–39
for marine turbine power
generation system, 42–46,
61–62
H-infinity Kalman Filter, 2, 9–10, 108,
114
stability of, 10
state estimation with, 122–123
Hypothesis testing, statistical, 10

I

Induction motor
five-phase, gas-compressors
actuated by

approximate linearization of,
118–122
dynamic model, 114–118
fault detection, 123–124
fault diagnosis for, 113–138
fault isolation, 124–126
simulation tests, 126–138
state estimation with H-infinity
Kalman Filter, 122–123
Inverters
three-phase, 1–2
approximate linearization, 6–8
dynamic model of, 2–6
faults affecting, 6
simulation tests, 12–16
statistical fault diagnosis with
H-infinity Kalman filter, 9–12
voltage source, 1

J

Jacobian matrix
gas compressors actuated by
five-phase induction motor,
119–121
marine turbine power generation
system, 43
three-phase inverters, 7–8

K

Kalman Filter, 6, 83, 84, 102, 172
derivative-free nonlinear, 86, 140,
178
state estimation using, 153–154
disturbance observer based on,
85–86, 178–180, 208–210
performance of, 88–92
H-infinity, 2, 9–10, 108, 114
stability of, 10
power grid sensors monitoring
with, 205–206
state estimator based on, 177–178
statistical fault diagnosis using,
155–157, 180–182
Kirchoff's voltage, 2

L

Lie algebra-based control
    for marine turbine power
        generation system, 47–50,
        62–67
Lie-Backlünd equivalence, 174
Linear approximation
    gas compressors actuated by
        five-phase induction motor,
        118–122
    state-space model of DC microgrid,
        102–104
    three-phase inverters, 6–8
Luenberger-type state-observers, 16
Lyapunov equation, 44–45

M

Marine turbine power generation
        system, 37–40
    differential flatness for, 84–85
    dynamic model, 39, 40–42
    fault diagnosis, 82–97
        fault detection, statistical tests
            for, 86–87
        fault isolation, statistical tests
            for, 87–88
        Kalman Filter-based disturbance
            observer, 85–86
        simulation tests, 88–97
    flatness-based control, 50–56
        in successive loops, 53–56
    fuzzy multi-model-based control,
        75–77, 82
    Lie algebra-based control for,
        47–50, 62–67
    multi-model fuzzy control for,
        59–61
    nonlinear optimal control for,
        42–46, 61–62
    PID control for, 78–81, 82
    simulation tests, 61–82
    sliding-mode control for, 56–59,
        67, 73–75
    state-space model, 41

Measurement update part
    H-infinity Kalman filter, 9, 105,
        122
    Kalman filter, 86, 178, 199
Mechanical part of PMLSM, 20
Microgrids, 98
    DC, electric power, *see* electric
        power DC microgrids
Model-based fault diagnosis, 1–16
    three-phase inverters, 1–2
        approximate linearization, 6–8
        dynamic model of, 2–6
        faults affecting, 6
        simulation tests, 12–16
Model-free fault diagnosis, 16–36
    Permanent Magnet Linear
        Synchronous Machine/Motor,
        16–19
Model Predictive Control (MPC),
        39
Moment of inertia, 41
Multi-model fuzzy control
    for marine turbine power
        generation system, 59–61

N

Neural networks, 17–18
    feed-forward, 21–23
    modeling of PMLSM with use of,
        21–26
    statistical fault diagnosis using,
        26–30
    using Gauss-Hermite activation
        functions, 23–26
    using 2D Hermite activation
        functions, 25–26
Nonlinear Model Predictive control
        (NMPC), 39
Nonlinear optimal control, 38–39
    for marine turbine power
        generation system, 42–46,
        61–62
Normalized error square (NES), 10, 87,
        106–107

P

Particle Filter, 16
Permanent Magnet Linear Synchronous
    Machine/Motor (PMLSM),
    16–19
  dynamic model, 19–21
  electrical part of, 20–21
  mechanical part of, 20
  modeling, with neural networks,
    21–26
  simulation tests, 30–36
  state-space model of, 21
  statistical fault diagnosis using
    neural network, 26–30
PID control
  for marine turbine power
    generation system, 78–81, 82
PMLSM, *see* Permanent Magnet Linear
    Synchronous Machine/Motor
    (PMLSM)
Positive definite symmetric matrix, 122
Positive semi-definite symmetric matrix,
    122
Power grid, 201, 203–204
  fault detection, 206–207
  fault isolation, 207
  sensors monitoring with use of
    Kalman Filter, 205–206
  simulation tests, 210–214
  three-area three-machine, model of,
    204–205
  usage of Kalman Filter as
    disturbance observer, 208–210
Pulse width modulation (PWM), 100
P-value test, 35

R

Riccati equation, 8, 39, 44, 122

S

Sigma Point Kalman Filter, 16
Simulation tests
  electric power DC microgrids,
    108–113

gas compressors actuated by
    five-phase induction motor,
    126–138
gas-turbine power generation units,
    157–167
marine turbine power generation
    system, 61–82, 88–97
Permanent Magnet Linear
    Synchronous Machine/Motor,
    30–36
power grid, 210–214
steam turbine power generation
    units, 182–190
three-phase inverters, 12–16
wind power generation systems,
    201, 202–203
Single-Machine Infinite Bus (SMIB)
    model, 42, 62, 141
Sliding-mode control, 39
  for marine turbine power
    generation system, 56–59, 67,
    73–75
Sliding-mode observers, 16
State estimation
  with H-infinity Kalman Filter,
    122–123
  with Kalman Filter, 198–200
  using derivative-free nonlinear
    Kalman Filter, 153–154
State-space model
  electric power DC microgrids, 99
    approximate linearization,
      102–104
  gas-turbine power generation units,
    146
  marine turbine power generation
    system, 41
  Permanent Magnet Linear
    Synchronous Machine/Motor,
    21
Statistical hypothesis testing, 10
Steam turbine power generation units
  differential flatness properties of,
    172–177

dynamic model of, 169–172
fault detection, 180–181
fault diagnosis for, 167–190
fault isolation, 181–182
faults and cyberattacks affecting,
    172
Kalman Filter-based state estimator
    for, 177–178
simulation tests, 182–190
Successive loops
    flatness-based control, for marine
        turbine power generation
        system, 53–56, 70–72, 82
Synchronous-generator power unit
    gas-turbine, *see* gas-turbine power
        generation units
    marine turbine, *see* marine turbine
        power generation system

T
Taylor series expansion, 2, 7, 9, 43
    first-order, 38, 113, 118
Three-phase inverters, 1–2
    approximate linearization,
        6–8
    dynamic model of, 2–6
    faults affecting, 6

simulation tests, 12–16
Time update part
    H-infinity Kalman filter, 9, 105,
        122
    Kalman filter, 86, 178, 199
Tracking error dynamics, 44

V
Voltage source inverters, 1

W
Wind power generation systems,
        191–192
    differential flatness properties of,
        196–197
    dynamic model, 192–195
    fault detection and isolation,
        200–201
    simulation tests, 201, 202–203
    state estimation with Kalman Filter,
        198–200
    transformation into input-output
        linearized form, 197–198

Z
Ziegler-Nichols technique, 82

For Product Safety Concerns and Information please contact our
EU representative GPSR@taylorandfrancis.com Taylor & Francis
Verlag GmbH, Kaufingerstraße 24, 80331 München, Germany